AF333460

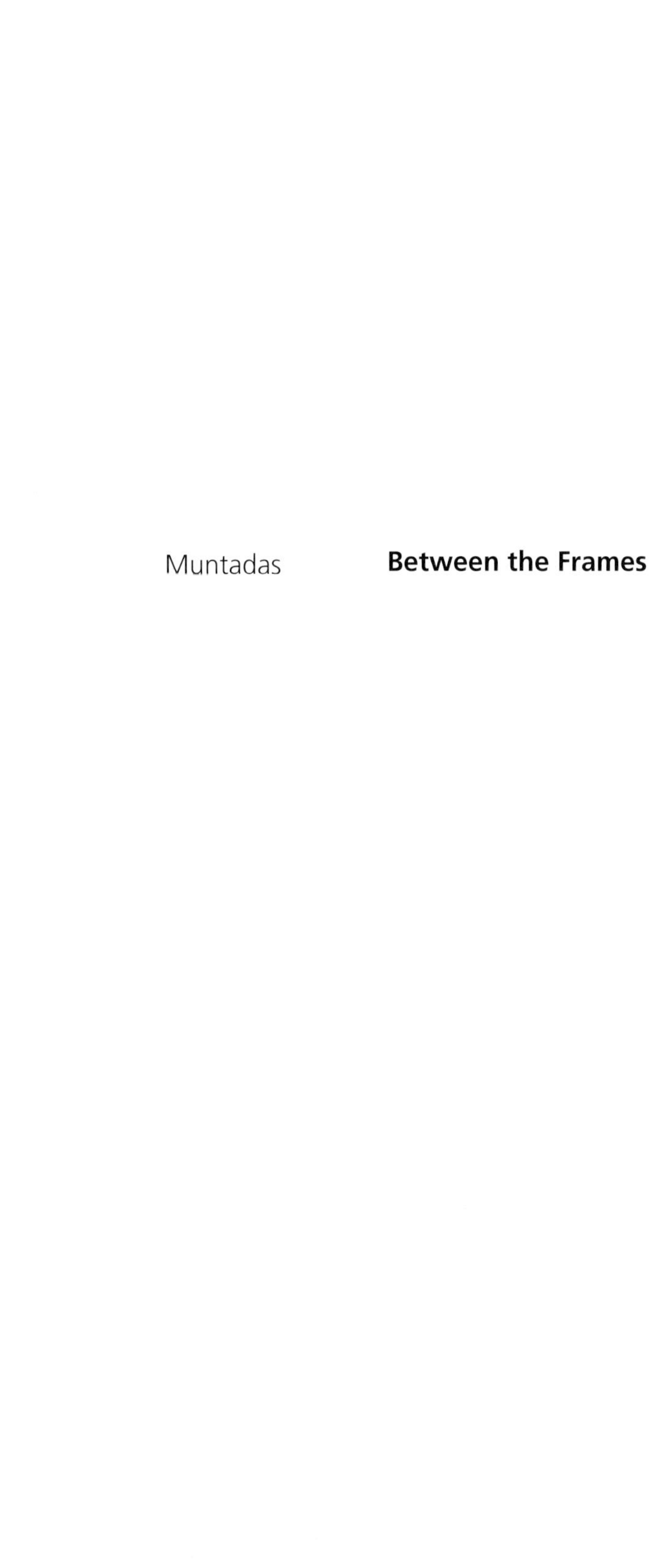

Muntadas

# Between the Frames

Interview Transcript

Wexner Center for the Arts
The Ohio State University

List Visual Arts Center
Massachusetts Institute of Technology

Published in conjunction with the exhibition
**Muntadas**
**Between the Frames: The Forum**

Wexner Center for the Arts
The Ohio State University
October 1–December 31, 1994

List Visual Arts Center
Massachusetts Institute of Technology
October 7–December 17, 1995

This exhibition has been made possible by a generous contribution from  the National Endowment for the Arts, a federal agency. Additional funding for the Wexner Center presentation has been provided by the Wexner Center Foundation and the Ohio Arts Council.

Transcription and translation of the interviews:
Muriel Ansenns, Caterina Borelli, James Morris, Daniela Noé, Joan Olivar, Simon Pleasance, Angel Shaw, John Sheplei, Carlota Subiros.

Graphic designer: Oliver Schmoldt
Copy editor: Ann Bremner



Published by the Wexner Center for the Arts, The Ohio State University, North High Street at 15th Avenue, Columbus OH 43210 and List Visual Arts Center, Massachusetts Institute of Technology, 20 Ames Street, E15-109, Cambridge MA 02139.

ISBN: 1-881390-08-X

# contents

**between the frames**

Art as part of our time, culture and society, shares and is affected by rules, structures and tics like other economic, political and social systems in our society.

*Between the Frames* is a series of eight chapters about people and institutions located between the artist and the audience.

**deal-er**
- to sell or distribute something as a
  business

**gal-lery**
- room or building devoted to the exhibition of
  works of art

•• an institution or business exhibiting or dealing
  in works of art

# 1. role

### Mary Boone

A gallery is involved with the dissemination of ideas. It's involved with presenting work, bringing ideas to public attention. And dealers are involved with selling pictures—two very different activities.

### Michel Durand-Dessert

First of all, a space for research; next, a permanent center for information about artists; and finally a commercial dimension. These remain fundamentally the same.

### Helene Winer

Strictly at our gallery we arrange shows and represent artists and exhibit and sell only that art. We don't have any back room activity or take things often on consignment from artists other than our own. So, it's definitely a gallery activity.

### Lucio Amelio

The gallery is the first catalyst of a certain creative energy that has collected around certain artists, certain people who wanted to say certain things. So, the gallery is a liaison between those who want to say things and those who receive and understand.

### Richard Kuhlenschmidt

At the same time that I sell art, I also work with artists, helping their careers along. So there's more than just dealing. I'm not just out there selling a product. It is more like working as an agent in certain ways.

### Holly Solomon

I am a dealer and I have a gallery where I show artists' work and try to sell the work. I try to see it as simply as I can because I tend to complicate things greatly.

### Joan de Muga

Of course the gallery has an informative role; it has a cultural role, if you will; but most of all its role is to create new collectors.

### Ronald Feldman

I think that galleries are taking risks. They are taking important risks. And some of the alternative spaces are also taking these risks, but galleries are taking the risks with their own funds. They can't get any kind of public funding. And I think that there are enough galleries that are doing this to give them, at this point, an important influence, which to me means the ability to get an audience to pay attention to serious art. It doesn't mean you can sell it, but it means that you can begin a dialogue with an audience with something that really is worthwhile.

### Ileana Sonnabend
Being an experimental gallery means that you have artists who do not sell as rapidly as others. And still I think that what they do is very interesting. They need more time because they make more demands on the collectors. You need to understand what they're trying to do. So, it takes more time, and we have these artists under stipend, which means they get a certain amount of money. Enough for them to go on working. Sometimes it takes a few years until we can sell. But up to now, it's been working out.

### Mario Diacono
My role as a gallery director is to be a parallel to radical thinking on art.

### Rosamund Felsen
I'm sure that all dealers would say that they have their own particular idea and philosophy of how a gallery should be, and I certainly have mine. And I have very strong feelings about that—not only the type of art that I show here and the artists I represent, but how it is presented to the public and how the artists are worked with. I think that a lot of my background in collaborating with artists at Gemini has formed my attitude and my philosophy of how I deal with the artists that I work with at the gallery. And I do think of it as a collaborative effort.

### Al Nodal
What's important about these artists' spaces is that they relate to the artist. They have the artist in mind more than other institutions. Ten, fifteen years ago, these artists' spaces or alternative spaces started producing art, producing programs. In a sense, they demonstrated what can be done. And now, all that sort of attitude, that style, that kind of taking risks at some level—all that has been sort of annexed by the museums as well.

### Joy Silverman
The galleries are taking care of some of the

younger artists and are taking care of sort of the financial needs and the business needs. We have more specific needs. We do projects where we challenge artists to do things that they wouldn't normally be doing with their gallery or to show their video work or their performance work. Or to put their work in a different kind of context, in a particular curated show or whatever. So, there's still a strong need for artists' spaces. Its just very different than what it was ten years ago.

### Marisa Díez

I am interested in presenting artists' first exhibitions and continuing with them. I think that a gallery must go with artists whose first works it has shown and then stay with them, follow their development.

### Leo Castelli

I don't consider myself a dealer, because what I and some other galleries similar to mine do goes well beyond just buying and selling paintings. We—these galleries and myself—are interested in finding new painters and then if we do, because we believe in them, we handle them, support them and do shows that we consider very important, of museum quality.

### Daniel Templon

A gallery is everything. You can't separate the cultural aspect and the economic. But at the same time there must be a balance in the choice of artists. You can't exhibit only very old artists or only very young ones, or only artists of one and the same tendency. There must be a general balance. When it comes to the choice of artists, it's obviously dictated first by personal taste; you only exhibit artists you like.

### Ivan Karp

There are probably thirty or forty of us in the art business who share some point of view. That is, we're interested in showing the best living talent. That's our main purpose. It's strenuous and some-times very exhausting and emotionally depleting, but also a very stimulating kind of activity. To see new artists all the time, to have relationships with the artists that exhibit with you, to deal with their problems of exhibitions and fame, to deal with their career problems—these are things that we do on a daily basis. It requires a great deal of patience and diplomacy and communicative powers. We have that

**12**

here. I know we do, and I think I share this particular outlook with a number of other dealers whom I
respect.

## 2. values

### Marian Goodman

I guess we need to believe in heroes, not in a grand
sense, but people who can change life and the
meaning of life. So that is a satisfaction with the
romantic part of being a dealer, and it's wonderful
to be involved with people who are so creative and
talented, but one also needs to make it work. So,
the financial aspect comes into it, as do the efforts
that a dealer can provide to help give an artist an
appropriate forum or make sure that people do see
the work and that the work can exist and survive
and succeed. So it is a combination.

### Ileana Sonnabend

I'm very pragmatic, and I have to be very much
involved with what I'm showing. I have to believe in
it very strongly. Since I believe in it, I show it and I
hope that other people will also get to believe in it.
It may be a little bit arrogant, but I think that what I
think is good, other people must think is good also.

### Joan de Muga

I don't believe in the purely romantic art dealer,
because to run a gallery successfully and be able to
present the things you enjoy, even if they don't sell
that well, you must be ready to present work that is
commercially successful so the gallery can function.
So if you mean to be serious in your work, you
ought to have a strong economic base created by
the art business. That allows you to get involved in
the artistic adventures you like and that you could
not embark on without having that economic base
that only the business can give you.

### Rosamund Felsen

I think initially I never really had to support myself
before, and I think when I first opened the gallery, I
was rather naive and I knew that it would be
difficult because I had heard for years how difficult
it is to have a gallery, particularly showing the kind
of work I was interested in. But none the less, I
thought, well, everyone says it takes anywhere from
two to five years before you can really start making
headway financially. So I just assumed that if I aimed

to keep my doors open that long, it would automatically happen. Well, it doesn't automatically happen. It takes a lot of work, a lot of energy. So, I think at the beginning perhaps my attitude was more romantic. I've learned with experience to become a little more practical and have a better understanding of the economics of the business.

### Helene Winer

Obviously, I would like to think that the economic realities don't influence me at all, but of course they do, and over time, they do so more and more because one just has to be pragmatic or realistic to a degree and also because of the phenomenal activity we have seen in the last five years, which is the entire time we have been open. It has been very, very active, with a lot of new collectors and a greater amount of art purchasing, and I guess this very active market just simply can't be ignored. It has influenced artists, it has influenced the art that they might choose to make, decisions that they might make. And, I presume, decisions that we make too, although I would prefer to think not so much.

### Lucio Amelio

In America, the gallery came into being with the precise intention of making money, dollars, big bucks, that is in order to sell objects, and also, but in rarer cases, like Leo Castelli, Sonnabend, in order to create a movement of ideas. The gallery here is more a center of discussion, first, and then a business. But there is a pretty big difference between the American gallery and the Italian one. The Italian gallery is still, in a sense, a little more romantic.

### Leo Castelli

The romantic side is much more important to me than the other side. The other side, actually, comes really way behind. The economic side has to be taken care of because there are, and there were, some dealers who are only romantic and have absolutely no talent for the practical side, for the pragmatic side. And those people just cannot survive. So in order to survive, you have to be practical. Museums get funds from all kinds of sources. The funds that we get, we have to get by selling paintings.

### Holly Solomon

So the demands are great. The emotional life is very great too, because it's not only your money at stake, but it's your emotions. You're dealing with people on a very personal level, and it's very hard because you are not family, you know. You are not in charge. I'm not their mother, their father, their brother, their sister. I am a dealer, and yet I am dealing in arenas that are very personal. It is also a business in which you have to be very discreet and know when information is available on a public level and when information is private. Because we are dealing with people's privacy.

## 3. market

### Daniel Templon

Having an exhibition today in a museum is not a sign of recognition. But recognition will come later, recognition will always come through the market. There's no other way to evaluate a work of art than to put a price tag on it. And that's how it's been ever since art has existed.

### Ivan Karp

The basic structure of our activity is to contribute to the visual arts culture. Of course, this is a very idealistic idea, and we persist in believing in it. However, we have to also survive. We have to hope, and we have to work at selling the things we have here. And we have some measure of success at that.

### Joy Silverman

The attitude right now in this country is very scary and very depressing, and more than ever, we have to stick with what we're doing and not loose faith in art itself and not constantly have to think of the marketplace. At least some of us have not to think about the marketplace and just think about presenting work.

### Ileana Sonnabend

And there must be a harmony between the price and a person's talent, between the price and the talent of the work. Acceleration is a very dangerous thing.

### Al Nodal

Everything that we do, in a sense, works towards increasing the market, because we are producing work, we are working with artists. A lot of what happens in our alternative spaces later on obtains a commercial value and gets sold. I don't think a lot of us—I mean at least when I was at the Washington Project for the Arts in Washington D.C. and even now in L.A.—worry about the art market. My whole economic situation is not related to it right now. So, I do not do it very consciously, but I know that, one way or another, everything that I do sort of supports the marketplace.

### Helene Winer

We, in particular, try to be rather careful about that. But then, when you realize that one of your artists is in huge demand and other comparable artists with comparable reputations—I mean, popularity—are getting very high prices; well, it is in a way cheating the artist, and ourselves to a degree, businesswise, to not put prices at a level that is similar to those of their peers. But we are very hesitant about over doing that, and very careful. And I am sorry that prices are high, as high as they are, and for certain artists it will—you know, like real estate—it will not be an error on the part of those paying those prices. But for other artists, it's really totally silly. Their careers, their importance, will not be able to support that kind of price, which means it will damage the longevity of their career, really.

### Leo Castelli

What really makes the prices go up is not what I do, but what the secondary market does, what other dealers do. They buy things from me or they buy them on the market and then, of course, they have to make them more expensive to make a profit. And then, little by little that jacks up the prices. Well, it is just an old law of demand and supply that plays an important role there.

### Ronald Feldman

If I know that a collector may not buy something until it has been confirmed by other sources in the art world, then sometimes that can be arranged. A very quick way, which has been used recently I think, is to buy a work of art at auction or to have a work placed there, whether by accident or just intentionally, and bid the price of that artwork up to a high price. And everyone says: We do not care what curators think. We do not care what critics think. The market is confirming that this is worth a lot of money. I think it is a dangerous road, personally, and I think it is totally artificial.

**Ivan Karp**
I will be quite candid with you about this: I don't often do this. What happens if you sell a painting to somebody based on its financial increase in value, and it turns out that it is not the kind of painting they like, and you have compelled them to acquire it for financial reasons—the painting is likely to come back and haunt you. Because the persons involved are not going to be happy with it. Paintings are a very, very strong substance, you know. They enter your life in a very vivid way. They occupy the arena of your daily existence, all the time. They reach out to you, and if you're not going to be comfortable with what you acquire, something peculiar is going to happen. Those paintings are going to be returning. So it is very, very rarely that I will say: This artist's situation is very progressive in the way of his prices. I will say that to people who ask me about it. If somebody says: Has this artist's situation improved dramati-cally in the last little while? And I will say: Yes, his price structure is progressive. We expect that next year the price may be twenty percent higher than it is now.

**Rodolphe Stadler**
I wonder also if one is not maybe somewhat a victim of the commercial system. I think there's been a tendency to use sales techniques for art that are pretty close to those used for consumer items.

## 4. network

*Muntadas*
*It is also related to a certain kind of market explosion—How do you analyze that?*

**Mary Boone**
I try not to think about it.

**Ileana Sonnabend**
There is no one-to-one relationship. It all has to go along together. Many people must be enthusiastic. Good collectors must buy. Museums and critics must be interested. And only when all of these conditions happen are you really secure.

### Joan de Muga

Often it's the galleries, rather than the artists, who dictate what the collector must be interested in.

### Ronald Feldman

I think sometimes there were false leads and false roads that occurred, but I think basically the galleries can have an important influence, but they cannot have it alone. There is a support structure which includes critics, curators, collectors, people who visit galleries. Other artists I think are the most important factor in creating taste, really. But a gallery can be a focus for all these activities, and I think because of that, it can have a very wide influence.

### Lucio Amelio

In America, and this is the amusing and interesting thing, there is a complicity in the art structure, which does not exist here: an almost daily complicity, I would say, that goes on second-by-second. You have the museum, the gallery, the artist…it is a "gang" that functions perfectly. Naturally, this perfect complicity, which exists in the structure of art in America, often creates a situation of market inflation as well.

### Leo Castelli

If my choice is good, then, of course, the fact that these artists are represented by my gallery is to my entire advantage. If the artists are not good, if the public does not accept them—and that happens pretty often—then there is absolutely nothing that they can do. I depend on some kind of mysterious consensus that gets established about a painter. I can't really impose a painter at all.

### Holly Solomon

And I think it is a very real thing. It is mysterious how somebody finds it. And I don't know if there's a real consensus. If it's really cutting edge, I think it remains so controversial and it doesn't please everybody.

### Ivan Karp

I don't think that people come together to agree that a certain artist should be launched in a certain way. I don't think it can be done quite that easily. But if a certain group of people representing members of the arts establishment—that means some museum people, a few critics, a couple of very strong dealers, maybe two or three strong collectors—if they agree that something should happen, then it is likely that some progress for particular artists can be made. And maybe that progress may not be meaningful or significant, but it will be made anyhow. So certain artists are likely to attract a great deal of attention that they do not deserve, while other artists who don't have particular personality traits may be left in the shadows.

## 5. fashion

### Daniel Templon

I consider fashion as an elevator that allows artists to "come into the spotlight" faster, to get known faster. If they're bad, they'll come down as fast as they went up. So fashion should not be condemned. Of course, fashion involves mistakes in appreciation, but that's too bad for those who make them. So, once again, fashion is beneficial since one can reach the stage faster, be perceived, seen, recognized much faster. It is a phenomenon of our time that applies to all fields. Why should it be any different in the field of contemporary art?

### Ileana Sonnabend

And I personally prefer to disregard movements and just start to make up my mind about who I think is solid, and sometimes it makes me be less "in vogue." I prefer not to be "in vogue," but to last.

### Ronald Feldman

Artists feel they should get rich very quickly. They should come out of school, and the world should be waiting for them, and they want to get rich very quickly. This I think is both good and bad. I don't mind if they're motivated, but at the same time, I think that their art is not really well thought out. It's more a strategy for finding acceptance rather than something that is introspective and important.

### Leo Castelli

Like every product—art, after all, is ultimately also a product—art is influenced by all kinds of currents. And fashion is really a good example which can be likened more closely than any other product— watches or radios or television sets—to what happens in art. There is an element of fashion there that plays an important role. But I do think that the serious galleries try as hard as they can to eliminate that factor. All of us—good galleries—try to find artists who will not only be fashionable but will really stay. But obviously we can't always do that.

### Holly Solomon

The nice thing about art is that it is a changing thing. It does not mean that it is fashionable. I hate to think in terms of fashion, because then I would close the door and I would sell clothes or something else. And it is not like show business, where you have a hit or they close the door. Good art needs time.

### Daniel Templon

I think it is the pace of the time. We must accept it as such. We needn't be nostalgic for those times when it took an artist six months to do a painting and where each square centimeter of the painting was something thought out and pondered. Once again speed should be accepted as an essential element of our time.

### Helene Winer

It is unfortunate to see this heavy focus on the commercial situation. But how can it not be when all you read about, and see and hear about, are these phenomenal success stories of very young artists? And obviously this boom situation is attractive to a great many people. And those artists who are truly serious, they probably want to know how that works too, and they should know. And they must make intelligent choices about how they want to proceed. I find it boring and unfortunate that there is this focus, but I still don't think that it is the fault of making that information available.

## 6. audience

### Michel Durand-Dessert

It's certain that there's also a desire among artists, as there is indirectly among gallery directors, to reach the largest possible public. That seems to me fundamental. I mean art is not in a ghetto, and so it always has its desires. But I believe there's a structure that is very specific to the visual arts, which properly requires two publics. Since we're a gallery that holds exhibitions, there's one public that comes and that can enjoy whatever we present absolutely free. Anybody can come into a gallery, look at the exhibition, and, if he wants, have the information we provide, and in that way completely consume the exhibition culturally. And this free consumption by a large public is only possible because, on the other hand, there's another kind of consumption, which is the fact of the work being bought by professionals who may be in museums and by collectors or others. So there are these two kinds of consumption, and it is very rare in cultural fields to be able to have these two kinds of consumption. When you go to the movies, you buy your ticket and watch the movie. If you go to a bookstore, you don't go there to read the books and go away. So there is for us this specific element of two different publics, one needs the other and both are absolutely indispensable.

## Marisa Díez

Attracting an audience from the street, a very "fresh" audience, and it is that audience that interests me for the artists. I mean the relation between artist and new audience, that audience with "new" eyes which appears one day and asks you: Who is this? and you explain. I am very interested in all of that.

## Ronald Feldman

The audience is growing tremendously, and that is very important, and particularly it is growing in New York. However, the fragility of the economic structure of taking risks and showing new art is the same.

## Richard Kuhlenschmidt

Art is very difficult to sell. It takes a certain sophistication even to understand the work, and especially here in California where the audience isn't that sophisticated. A piece of art can be extremely well received in New York, but if the people out here maybe aren't familiar with it and with the concepts behind the work, they're not going to buy it. It has been very difficult for me because I do show a lot of artists from New York who are more or less conceptually oriented—at least their background or the history of their work comes from conceptualism. So it doesn't serve as a decorative thing to fit over the couch.

## Marian Goodman

One thing that I am concerned about in terms of the art world in general is that it is often the case that the art that is easiest to like becomes popular first, and the art that perhaps is a little less accessible or less easy to see and grasp immediately has a more difficult time getting accepted. I guess it has always been that way.

### Ronald Feldman

The way to survive the difficult recognition of some work is the way we survive: that is, not to count on the fact that you can sell your exhibition nor select it because of that. But you are going to have to be selling other known works, by twentieth century masters or older things; you have to know that kind of material. You are going to have to deal in it, or you are going to have to do some publishing. You're going to have to have some source of income that you can rely on, some real business within the art world or outside it that is your self-funding source, to carry on the activities as you think they really should be conducted, rather than making the wrong choices and showing things that you know are immediately saleable, or hoping they're immediately saleable and not making the difficult choices of what's really the best art at any particular time.

### Al Nodal

I do not relate to this whole field in terms of how it relates to the artwork being produced. I mean, artwork changes. Everything changes and you sort of keep changing the way you relate to art, the way you service the field in different ways, and things change.

### Holly Solomon

If we have any power, our power lies in the fact that we are independent. We don't have to follow the hordes. We don't have to show what is hot. We really have to look to our own resources and do what we believe in because I am not so smart as to be able to say: This show will sell out.

### Ivan Karp

The art community within itself feels like a very vibrant energetic community, which is what it is, but it's a very small portion of the American art scene, of the American scene in general. The art community is a small community. It is an intense community, and it's viable and strong within itself, but people are beginning to think that it dominates American life, which it doesn't. It is important to understand the role that we play. It has a significant role in the growth of American culture, that's for sure, but it is a small community. Everybody knows everybody and people have broken up into various factions. Each of these factions has very strong convictions and feelings about what art should be, and so it breaks down into facets of political conviction— political conviction within the artistic context. The politics of art are the same as politics in all the arts or in all walks of life. In music and theater and in literature and in business, there are political factions: certain people get together and they agree about something.

## Joy Silverman

To trust in artists who are not commercially viable at all and might never be, and give them a chance to do whatever the hell they want to do. I think that is still exciting. I think it is still exciting to give an artist a chance to fail, because you can't do that right now in the marketplace. This is going to sound like a throwback, sort of idealistic, but somebody's still got to keep that sensibility, that pureness in the art world: art is more important than all the other stuff. I think artists' spaces are among the few institutions in the art world that still do that. And it might not be trendy right now to have that attitude, but I believe in it. And that is the reason that I am still in this field and really committed to it, because it is pure. It is still pure. It is not based on what is selling. It is not based on making an artist's career. It is not based on making a curator's career. It is not based on any of those things. It is not based on making a collector's collection more important. It is based on solely what art is about, what is happening right now today, and the way we can just show that in the purest form.

## Glenn Lewis

I think that artist-run centers and alternative spaces are a counter force to a lot of that commercial gallery activity. Well, it will be interesting to see just how that all works out, if art will be seen as more of a business, if you like—if it becomes more and more business like, more and more commercial—or whether it will become less commercial, more to do with the idea of art and what art is in its essence rather than what it is as a commercial product.

chapter 2  the collectors

**col-lec-tor**
- one who collects

- one who makes a practice of collecting objects

### Herman Daled

In the beginning when I started—I have to use this word—to "collect" or when, inevitably, I started acquiring works of art, it had a meaning. I think it had, for me at least, a very profound meaning. It was a way of supporting the endeavors of artists who looked to me as if they were producing interesting work. And I would say that the ownership aspect was for me completely secondary. And I think it may have been for me a way of getting high and being part of something, and for me the essential aspect was participating in something that was in the making. At that time, it was very easy because it had to do with men and women who, as we always say, were my own age or a little younger. While now, later, it becomes more difficult because, in fact, I think that the contemporary manifestations are obviously in the hands of younger people, and inevitably there is a generation gap. But worst of all is having a label on your back that says "Collector" with a capital C.

### Robert Rowan

I have collected for a long time. I collect things that are going on now—young people. I have looked at a lot of young people, and I think the energy may be here instead of the east. I used to collect on the East Coast. It seems to be that California, particularly L.A. and San Francisco—the best younger artists seem to be here to me. I collect what I like. Why I like things is much more complicated, but I am just saying: I collect what I like, and I collect with the idea of giving. I am not without the idea of collecting something which is good enough for museums. Okay, so then I try to give—to give, to lend and give what museums want.

### Eric and Sylvie Boissonas

I am very visual. Some people, you know, are auditory, like my husband, cerebral, but me, I'm very visual, so whatever is around me has always played a big part. So in our family—my sister Dominique will tell you more or less the same thing—but we had a grandmother who was very open to those things, even to things completely avant-garde in her time. My mother was very much so, and we always saw not paintings of value, because at that time it wasn't so much in the minds of bourgeois Protestants to spend money for things like that, but some very beautiful reproductions in black and white. What part that played, I don't know. We always went to see exhibitions, and I must say one thing, that it was our stay in America that really got us involved in the field. We lived in the United States from '46 to '60. After living for a short while in Texas, we moved to Connecticut. I would go several times a week to attend courses in art history at Columbia, especially those of Schapiro. All the collections were there, all of them. And I must say that my sister too, Dominique de Menil, although she was living in Texas, she would often come to New York; she had a passion for art and bought works. All that had created a climate.

### Giuseppe Panza di Biumo

I believe my task is to try to find what is important in the art world very early because I believe the best function of the collector is just to look and understand at the beginning what is good in the art which is underway. I believe a collector can be useful if he has this attitude. The task of the collector is just to choose the best work between the many works an artist makes.

### Marcia Weisman

Basically, I love art and as a result I think we all have pretty much the same feelings—what we love we want to share with others. We want them to love it too, maybe to give validation to what we do. And maybe I wanted people to love art because I love it. And then again, I had many personal contacts with artists and in many instances I didn't think artists got all the breaks they should have, and I was interested in helping them. But not as a patron of the arts, because I don't think I want to be patronizing artists. I think I am also a volunteer, and I think a trustee of a museum is one of the highest forms of volunteerism there is. I do not know whether a picture is good or bad unless I see another picture next to it. So when I see another picture next to it, I know if it is better or worse. And I look a whole lot. I look, and I look, and I look. I think I have learned from looking at, not from reading about, art—anymore than you can read about music: you have to listen to it. And you select what you like the best, based on everything you have seen. And if you have looked enough, one day you will see something, and I do not know how to express it in good English, but all I know is it gets you right here. And you know that everything you have seen you bring to it, and when you see that with everything you have seen brought to it, it does it, it happens, and that is yours, and I have to have it—I should say that's mine, and I have to have it. Obviously, I can not have everything I want, although I think I am entitled. I would like to have it. I think everything that has been chosen…everything that you see and that I have collected through the years are works of art that I have fallen in love with and that I love, I want. And I have seen many other works by that particular artist that I bring to it. And I think you only learn by comparing one thing with another. It is no different from music. It is no different from the clothes you buy, the shoes you wear, the furniture you buy—you name it—the woman you love. Everything is compared to the others you have known of that particular category. It is not something that must be looked upon as special in the sense that you have to have a certain mysticism to understand art. I find too many people are intimidated by art. They say: I do not understand it. What is to understand? You like it or you don't. You love it or you don't. And after you have loved it and you want it, you want to live with it, and after you have that feeling, that to me is the appropriate time to learn about it. To learn who the artist is, where he painted, why he painted, how he painted. That's all very important, but it comes after the real thing, after that first instinctive reaction, as I see it. Now, I know that there are people who collect differently than I do.

**Fernando Vijande**
I think of myself more as a collector than as a dealer, and also because usually the pieces I sell I have bought myself; they are actually mine. When I sell a piece I always get upset. I love to sell them, especially to a good collection, because it is important for the artist to be well located, but if it is such a good collection you want to sell the best piece so that the artist is best represented. And I find that very hard. I mean, it is painful for me. That's why I usually don't like to sell, I prefer to have other people doing it in the gallery.

**Bob Calle**
I felt a certain need to commit myself. Which means it is all very well to say: Now I like this; I do not like that. Or to criticize this or that painter, or

to criticize this or that attitude, but I had to
participate. Well, I'm not a creator, and if you want
to participate somewhat in creation, what you have
to do, in fact, is buy works by young artists,
commit yourself, take a risk, even if it is a financial
risk. But it is not only a financial risk, it is an
intellectual risk. Even people who have a lot of
money must choose among several painters. There
are so many things being offered so we choose
one. We choose to buy this or that painter, this or
that work of the painter. Even if we like a painter
very much, there is a choice that has to be made in
his work: this or that painting. It is a way of
participating—I would not say in the act of
creation, but in the development, in the end—by
helping, by encouraging the painter, since for the
painter it is always an encouragement when we
buy something from him. And also to make in
some way a final accounting of the acquisitions
we've made, also of the ones we haven't made, by
saying: I hesitated, I did not understand this or that
painter at a certain time who already had quality.
But we did not understand, and so we missed
things.

**Acey and Bill Wolgin**
We have a very eclectic collection, and mainly it is
people that we know or we've met or we've spent
time with and artists whom we like. Maybe that is
not so good. I am not sure, you know, but that is
the type of collection that we have. It is things that
we like and things that we love and in association
with the people who have made them. There are
some people who have lots of money, and they will
get an adviser—who is usually a gallery person—
who will advise them to buy this piece of art
because this piece of art right now is pretty cheap
and it will appreciate, etc. and it is a good investment. Or buy this piece of art because it is well
known and it is already expensive, you know, but
will probably, when inflation goes up, be worth
more money. That is a different type of collector,
and I think that there are all kinds of collectors.

**Gianni Rampa**
All these things, these few things that I have,
which are not that many, but that I dearly keep and
love, are in fact the works of artists whose
existence and thought I read about. In America,
great collections that go into museums come from

private people. It is the private person, the big, rich man, the industrialist who, at a certain point, does collect partly out of vanity—and I mean this in a good sense—and partly because he knows very well that the works will produce an economic reward.

**Isabel de Pedro**
We began by having the pictures that we like hanging on the wall, and then, as years have been going by, you start getting interested in other things, you start getting interested in theater, in music. You start understanding that it is not so important to have a picture hanging, but that it is much more important to live closer to art, to participate, to see many things.

**Rafael Tous**
After all, it has also been a preservation of this whole world of art all around us, which otherwise would disappear. I mean, this kind of collector of contemporary art—in this country there are very few, you can count them on your hand—and so this has also been a little bit a preservation. Almost everything has been bought because of a friend-ship with the artists, because of a dialogue with them, and never from thinking: This picture, or this piece, or this film, or this video is going to be worth more money in the future. But because we liked them at a certain moment and there was a dialogue with the people we bought them from, and because of the relationship with these artists. I mean, it has not been just a matter of speculation, of making an art collection that is going to have a certain value, but because of our friendship with the artists and for the preservation of a culture that public institutions were neglecting. So we had to do it because we had the means.

**Toshio Hara**
Collector…I think there are many kinds of collec-tors, and the role that collectors can play depends very much on why they collect. I think my case, as you say, my motivations are private and have very much to do with my own spirit and my own soul. So unless I am satisfied with what I am doing…Just trying to be well organized so I can serve the public, I think that comes as a result of my collection, the fact that my collection is open to the public in the form of a museum as a result of

**30**

having the motivation to collect. But if you ask
what is the role of the collectors, I think there are
many effects which result from their activities. So,
being public or showing works to the public or
organizing certain activities: these things come as a
result of your activities. And I think for collectors
what is important is why they collect. So many
people, for instance, collect for the sake of money,
for many other reasons, I do not know. It is very
private in a way. I think it is rather dangerous to
generalize the result alone, saying that collectors
should do this and that. I think most collectors are
doing it with quite different motivations.

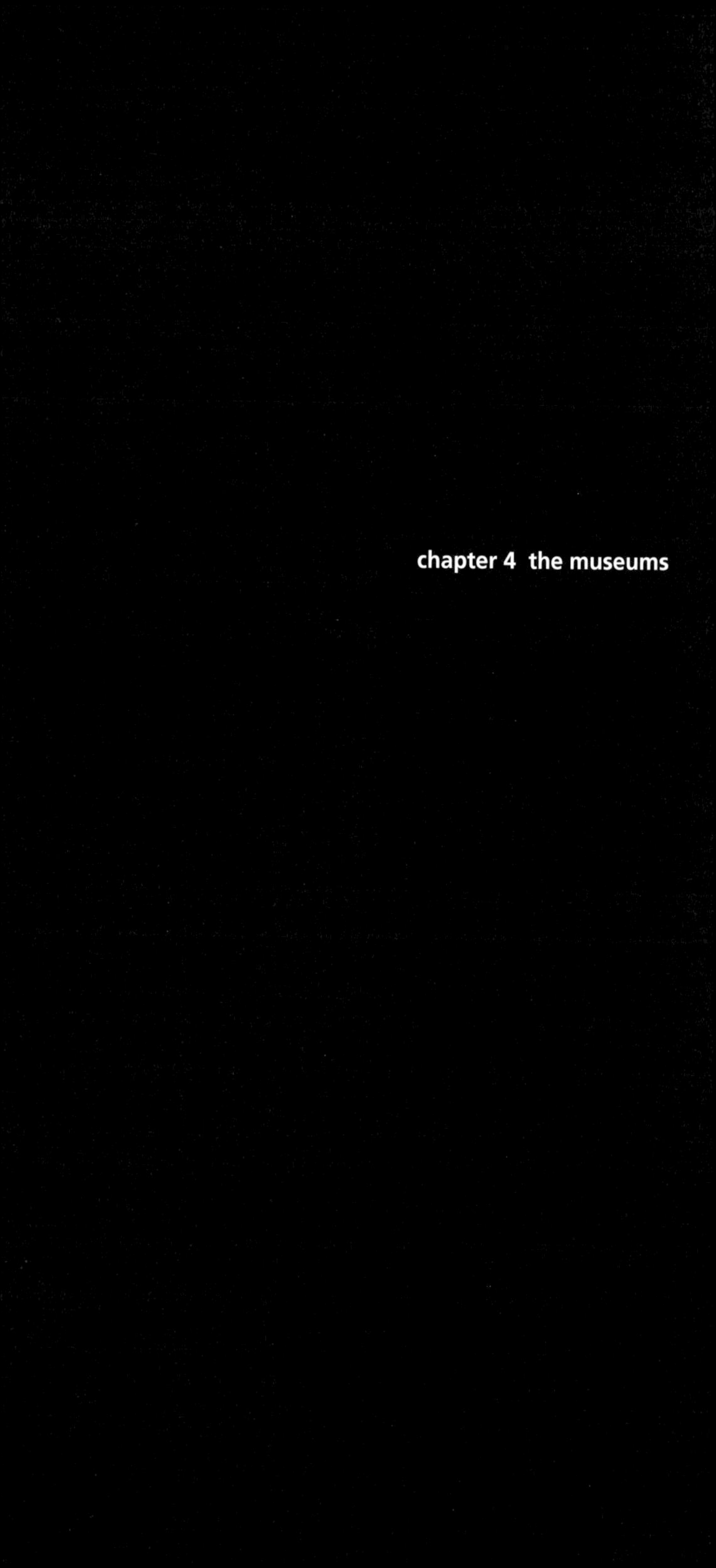


chapter 4  the museums

**mu·se·um**
- an institution devoted to the procurement, care
and display of objects of lasting interest or value

# 1. role

### Harald Szeeman

I consider the museum mainly as a place where you still can experience these fragile products or creations: the only place. And this is very important because all that is utopian and not yet discovered by a majority, you can show it when it is only understood by two people. So, you have these places where, really, you can have the alternative to the mass media influence. After I curated *Documenta 5*, I decided to be no longer connected with any institution. And so I called my intentions: "Agency for spiritual guest work in the service of the visualization of a museum of obsessions." And of course, an obsession can have a visual character, a written character, a behavioral character. I mean, it is, in a way, to declare the universe as your museum—and that is still the point.

### Anne D'Harnoncourt

The great role of the museum is to bring together people and works of art. And to create that encounter which is so important both to the life of a work of art and to the life of people. Beyond that you can go in almost any direction. I mean, museums are really kind of coalitions between a broad public, art historians, artists, collectors, people who aren't so interested in art and suddenly they come to a museum and get very excited…There are the professionals within the museum. I'm always fascinated by the modern museum, which seems to me to take on a great many different kinds of roles: at one level, it is an educator; at another level it is a place for performances or for contemporary artists; at another level it presents a vast picture of the history of art to the public. But above all, museums are places where works of art and people encounter each other.

### Josep Ainaud de Lasarte

The aim of a museum could, in a way, be considered a contradiction. The normal situation of things is to be ephemeral, to have a life: to be born, live and die. The museum tries to make everlasting what is not always made to endure. And this of course creates a certain polemic. And there's yet another contradiction, maybe not as obvious: the museum aims to do two things that, as I see it, have their own internal contradiction, which must be surmounted so that they work out. On the one hand, the museum must preserve the cultural heritage of a given country, and—in this sense—the cultural identity of humanity. At the same time, it has to show and communicate this heritage. But often, to communicate implies a partial destruction. I mean, it is obviously easier to keep something in a safe or in the dark rather than within arm's reach. To find a balance between the need to preserve and the need to communicate is not easy, but it is one of the specific goals of the museum, the one that gives it its specific nature, that makes it different from other institutions.

## Thomas Messer

As director of a museum of twentieth-century art, I see our function as one of setting visual standards, of acquiring, presenting and making accessible to the greatest number of people works that are exemplary in their formal and visual perfection. That is the underlying purpose, it seems to me, accompanying features that have to do with didactic presentations, with research and publica-tions, with exhibition programs. But all of these are simply different methods of gathering within a museum works that, through their excellence and their importance, can convey the essence of visual perfection.

## Johannes Cladders

The museum, I think, is a medium. A medium: like a paper where you can read art critics, like a book on art, like a film on art and so on. It's a medium too, but I think it is a special medium, because all the other mediums I mentioned can't deal with original works of art. They can show only the reproduction, and I think it is a function of a museum to give priority to this aspect: to be able to show the original work of art.

## Marcia Tucker

This museum is devoted to the investigation of the art and the ideas of our own time. That may mean things that have happened very, very recently. It may mean things that have happened twenty years ago that have deeply affected what's going on today.

## Jean-Louis Froment

The question that I really could ask myself today is: What is the meaning of a museum at the end of the twentieth century, getting ready for the twenty-first century, when for example, the human race is threatened? When, for example, there are ideologies in which we, the people of my generation, believed for a very long time, which are now overturned so easily? When, for example, the world of art, the world of creation has been threatened by the world of production? So there are questions about what I'm doing here in Bordeaux, of course. A whole set of questions, which somewhat contradict each other, because I myself am looking for and don't have an absolute answer. I'm looking. So here's what I want to tell you: I see my work today as really the work of someone who is working in an observatory, in a laboratory, who absolutely rejects the idea that there can only be one way for art, for man. Right? And therefore, someone who proposes a multiple meaning for one proposition that would be art.

### Wolfgang Becker

We try to hold the position in the museum of what was in old times called the "arbiter elegantiarum": the institution which constantly criticizes, controls and discusses aesthetic validities, not only in art circles, but in design and furniture circles, and whatever you can think about. That it is a kind of judge who renews every year or every month his aesthetic judgements based on the experiences he acquires.

### David Ross

I think the notion of the museum as something with some huge prehistory is our own invention or need to make even more solid an institution whose hundred years or so of activity has provided a certain basis for the way we work today. After all, the museum is a nineteenth-century invention that follows the decline of the royal household as the repository of works of art and follows the growth of public education as a kind of norm and basic right, a basic concern of governments.

### María Corral

I think that museums are still the intermediaries between the artist and the audience. And I think that even if they have now become cultural shows, rather than spaces for culture, we must keep on fighting so that they are spaces for culture. I think that the role of the museum nowadays—and its role needs to be expanded from the nineteenth-century definition—is that the museum must be a place for provocation, where art does not end up but where it also begins.

### Rudi Fuchs
In one sense, a museum is for artists and for nobody else. I mean, the public can come in and witness the fact that there's an artist in a museum showing his work. But all circumstances within the museum should be organized in such a way that they enable the artist to show his works. That is one point. The other point is that the museum, together with other institutions and artists and everybody in the whole field, is weaving this fabric which is contemporary history.

### Pontus Hulten
In the older school, the curator—museum person— was essentially standing on the side of the public looking at art with the same optic as the public, while at Sondberg we were looking with the eyes of the artists.

### Margit Rowell
There are two ways of doing it. One is to think very hard about what you are showing in the contem- porary area and think about whether it has a certain kind of significance in the present day context, and be very selective. The other is just to sort of show everything and say: This is what is happening in the world today. I happen to think that it's a pity to do the second in a museum. I think that the people who come to museums, many of them are the same people who go to galleries, and I think that the galleries are showing what is happening today. And that the museum should do something slightly different. I feel that the museum is very important for the greater public, and I feel that the museum is very impor- tant for artists. In all the exhibitions that I have done, I think in the back of my mind I was saying I was doing them for the artists' community, in that I feel that good exhibitions generate better art, so to speak. And that is why, again, I am very selective, and I may be entirely wrong sometimes in my selections, but I think that that is preferable to just showing everything.

### Jean Christophe Amman
As for me, for sixteen years now I have been

setting up exhibitions, and basically that is where I see my role, making it possible for artists to do something in the space, and I do not think I am going to change my ideas on that. Of course, there are epochs or periods when art is more exciting. I think that toward the end of the sixties and the early seventies there was a general upheaval. And then, of course, the public, the artists reacted differently, but that was rather a question of how they perceived the museum. But basically, I have always conceived it as a place of research or a place of work.

## Richard Koshalek

I think museums are at a very interesting stage in their history. I think it is a very critical stage. And I think what is desperately needed in museums now is creative leadership, and I think what is needed is people who are willing to do things differently within the museum context. And I think that one of the reasons that there has been this tremendous interest in Los Angeles—in this museum in Los Angeles—is that there is an interest world wide in museums doing something differently and museums taking a few more risks, museums being more interested in sort of thinking about how they present the work of contemporary artists, thinking about how the public comes in contact with it and what is the best way to make these presentations and to think about this work. And so I think we are going to see, in the next ten years, tremendous change within museums. Museums are very complicated organizations. You have many different constituencies. You have the artist constituency. You have the trustee constituency. You have the "funding" sort of constituency: corporations, foundations, individuals who give money to an institution to support it. You have got the membership, you have got the volunteer constituency. You have got the political constituency—for us the city of Los Angeles. Bringing all those forces, all those interests, together and producing an institution that can be creative is the biggest challenge I think we're facing in the next ten years. To do that it's going to demand and there's going to be a need for a different kind of museum person. One who is much more general, much more willing to sort of tie all these different sorts of individual forces together, to make a situation happen that's right for a specific city. And I think museums are going to become

much more sort of regional, in a sense. I think
they're going to develop in a way in which they
draw strength from the region they're located in.
That doesn't mean that they don't show work by
artists from around the world, but a museum in Los
Angeles should be very, very different say from a
museum in New York City or from a museum in
Minneapolis. It should have its distinctive qualities,
and it should have its sort of individual personality,
and I think that's what's going to start developing
over the next ten years.

**Marc Schepps**
It is a very complex role, and I don't know if we can
say that every art museum, in every place, has
exactly the same role today. I may have certain
ideas about what a museum or a museum of
modern art could or should be in this or that place,
but I think it's very important to understand the
place where you are, to be imbued with that
reality, and from that reality, to build an imaginary
museum adapted to a specific reality.

**Diane Shamash**
I think that museums are undergoing a tremendous
change. I think the traditional role of the museum
was almost a cataloguing of objects and protecting
and interpreting objects throughout a long history.
I see museums now entering into a much broader
kind of public domain, particularly since the sixties
and looking at how many contemporary arts
museums have developed. I think it is phenomenal,
particularly in the contemporary world, that
museums are kind of taking on a much more
public role.

**Dominique Bozo**
An institution that is heir to the Kunsthalle concept
and the art of the sixties, which means no more
art, ephemeral art, etc. And also has a little to do
with the idea we had about local cultural centers,
"maisons de la culture," which have been after all
very, very present in France in the sixties and
seventies. It seems to me that we should preserve
this wealth of relations with the public and of
freedom and possibilities for action. I think that for
the twentieth century we have really reached the
time of the museum. Good. These are things we
wouldn't have dared to say a few years ago
because we would have looked like horrible

reactionaries. Besides, being a curator has a connotation of being stuck in the past. On the contrary, I think that the great institutions and great historical museums where the artists and the public can come can only serve as a basis for living art.

### Kathy Halbreich
I think as art and fashion become closely aligned, the market becomes a major influence. I think that this is the age of the market. There are more people who want to buy art. There are more people making art and there is more room to manipulate the buyer and the maker. I think institutions which are nonprofit, as this one is, have a responsibility not to muck around in those waters.

### Yves Gevaert
I find that at present there is such collusion between the museums, the galleries, and the critics: their turfs aren't defined. I think that up to the seventies each turf was clearly defined and everyone played his role with precision and respected everyone else. And now we step on each other's feet.

### Gary Garrels
In the eighties we saw a jousting between the marketplace where an evaluation was taking place in terms of a monetary index. Collectors competing with each other to get the works of certain artists and driving those values up. The art market is one of the last "laissez-faire" markets where if there is a demand then the price can just keep rising until you reach a ceiling where the demand is either saturated or won't increase—I mean the price paid won't increase. And it was frenetic. It was irrational. It was a boom. That is a classic pattern, whether it was tulips in Holland in the seventeenth century or whether…I mean there have been many instances of it. And I think we saw a boom period where there was kind of an irrational drive in the contemporary art field, probably because it was a field that was very undervalued for a long time. It became glamorous. It became a kind of system of "entrée." Museums have also functioned that way for a long time too. It was a way to clean up dirty money. A way for people with "new money" to integrate themselves with people who had "old money." I mean that has been a classic pattern for decades. I think it attracted a lot of people who were much more interested in the social aspects of it than they

were in the artist's intent, than they were in the
ideas, in the objects, whether conceptual, aesthetic
or sociological. I also think for some reason there
was a lack of a kind of curatorial advocacy or
critique, at that point. I don't know why—I mean
there are historical cycles. There is a dialectic at
work, and there was a group, a generation of
people involved with museums, who had been
there probably for a long time and were just not as
responsive or engaged with what was going on in
very contemporary work, and there was a group of
very avaricious collectors. I mean there were
collectors whose entire lives were spent going to
other museums, going to galleries, going to studios.
There were very few curators who were functioning
in that same dedicated kind of role.

### Fumio Nanjo

I think we should make a more precise difference
between several places. Like in Germany, they have
a Kunsthalle and a Kunstverein in each big city, and
their role is different. If we consider those places in
Japan—Saison, for example, or other so-called
museums in Japan—they may actually not be
museums in the context of the west. they are a
Kunsthalle type of place: you don't have collections.
And sometimes they sell works: they could be a
commercial gallery. So it's all mixed, and the term
museum is used for anything in Japan. That is very
confusing. But for Japanese people it is not really
confusing: they think that it is like that. But for you,
it may be confusing.

### Richard Amstrong

People who are interested in generating the
exchange of money use the museum as a lever. And
it is very much like…the church saying: You are
saved. It is very helpful for them to say that there
are "x" number of converts in the preceding year,
and they use that as a marketing tool. So I think the
museum endorsing a certain kind of art or artist is
used as a marketing tool. But that is a fact of life.
The Whitney Museum is not affecting the market in
a major way as far as I can tell. Especially at this
moment, the collectors really have all that power,
which is really where it should be.

### Jean-Hubert Martin

I prefer the European model because I think that
public money is a guarantee of freedom, more so

**41**

than private money. I mean by this that we all know some American museums that are controlled by a few trustees, and sometimes by just one person; and that consequently these museums are obliged to hold exhibitions of artists collected by the trustees. And we know that if we compare in contemporary art—I mean in the avant-garde—what happened in the European museums and in the American ones in the sixties and seventies, it is obvious that a lot more was happening in the European museums. That is clear. When it comes to collections, let us say, the problem of relations with the authorities or the people who give money, I think that both Europeans and Americans have to deal with problems that are pretty much the same. Well, in France there may be political problems that get in the way of the cultural administration. Similarly the Museum of Modern Art in New York could have problems with Mr. Rockefeller, who may decide that such and such an artist doesn't suit him and he'll let it be known, and he'll let it be known financially. So I think that these problems of getting along with the authorities who hold the purse strings exist one way or another whether the museum is public or private.

**Martin Kunz**
I think at the moment the museums in general, including myself here, don't use their potential. I think we are too weak in relation to the mechanism of art.

**María Corral**
Culture started to greatly interest politicians in a new way. In the past, politicians with a good culture might have decided to create a museum, a concert hall, an orchestra—not any more. Nowadays, culture is used for political purposes or business purposes. And this is a big problem since people organizing exhibitions today are people who might have nothing to do with the museums, with the academy, with the university, but are people who simply give money. They force you to put on exhibitions that are spectacles, which brings us to what we were saying before.

**Johannes Cladders**
They can decide on the money that I have, and they are right—I think—but they cannot decide on questions of art. There are no specialists, they know too little about art.

**Kaspar König**
Culturally and historically museums are very
significant institutions, and I do believe that
museums play a very significant role. I am not in
favor of these kind of populist, generalized
museums. I like the generalized museums of the
nineteenth century, you know, which have a
collection of anthropology and medieval art and
contemporary art, like say, the museum in
Darmstadt for instance, with its magnificent Beuys
block. A museum is definitely a cultural-political
tool of self-reflection and focusing on the future.
And I think this is what makes art so interesting:
each decade you have a different view of what the
meaning is of a certain kind of work of art. And if it
is not from a point of view of today looking back at
history, then it would be rather a dead issue, and
instead we find out this is still very vital and not a
dead issue at all. What I would like to see is more
local focus and a kind of global quality involved.
You have a lot of museums with collections,
contemporary art collections, which more or less are
all alike rather than being different, and, again, that
is also a rather negative situation. Thank God there
are a few exceptions, but there are very few. I do
feel that the museum should redefine itself, and all
the exhibitions and various kinds of strategies I have
developed in practical terms are always in the
dialectical relationship to the museum: they are not
against the museum, but they always take the
museum as a factor. It is good to have different
models, but each one belongs to a particular
culture. I think I am rather skeptical if we adopt a
kind of American method where the interdepen-
dence between collectors, trustees, galleries and so
on becomes so strong. And I am afraid whenever I
see colleague curators in America: there are very
few curators who have any kind of inner strength
because they are so dependent on a kind of social
circuit of collectors, trustees, art dealers and so on
that it is not productive for the future.

## 2. architecture

**Rudi Fuchs**
How you install a painting or a sculpture is something which I cannot answer in general terms,
because for me, it always depends on the particular
painting, and in that sense, I always have said—and
I will say it again—that I think making an exhibition
is a craft, it involves "craftsmanship," and is not an

intellectual work. It is something you do with your eyes and with your hands. I mean you cannot figure it out, you cannot think it. You can never think an exhibition. You have to actually make it and put it on the wall and at that moment it comes together. I think architecture is overestimated. In my view, rooms should be simple, for the moment. I am not saying that we should always make exhibitions in simple, white rooms, but what is going on now in architecture to me is too rococo.

### Carmen Giménez

You must be able to recognize the art and the artist, not the building or the curator. The focus has to be on art and not on the hand of the architect. I believe this is the ultimate goal. And, in ninety percent of the cases this is not what happens.

### Harald Szeeman

Always use architecture as it is, don't change too much. But I know that sometimes it is difficult. You see all these museums now, and sometimes the result is very good, I think. On one hand you mentioned Mönchengladbach, the Kleblat-Trefle space: that is a very good result, but downstairs with Beuys it's not a good result, and that is more Hollein's entrance—he did it in many tourist offices, with palm trees. So you see, I would really much rather take an anonymous young guy and tell him what the museum needs. Anyway, I am for museums: just walls, upper light, and, optimally, the possibility to darken the spaces.

### Jean-Hubert Martin

I am completely in favor of a functional architecture that could have many meanings, it could mean many things, but above all it should be a modest architecture. That is to say, it should not compete too much with the works of the artists. And that, as you suggested in what you just said, is very rare. I see three reasons for this. Today many cities or countries want to turn museums into prestigious buildings: consequently they need an architect and his signature. For an architect, a museum may be the last building where he can really create an "oeuvre" in the traditional sense of the word. And in the third place, there is this paradox: that museum curators and directors obviously tend— since they believe in creation—to look to an architect as a creator. Which means they accept his

fantasies, what comes from his imagination. And
by doing so, they hobble themselves and saddle
themselves with difficulties that are often enormous, because the architect does indeed create a
"piece," almost a work of art, and later on it will
be very difficult to place works of art in it because
sometimes the architecture will overpower it.

### Yona Fischer

Today I think that my wholly personal preferences
are rather for much more limited spaces. I used to
believe strongly in large spaces that could be used
endlessly like the new—well, not so new—gallery
of the Zurich Museum, for example, or like the
Whitney, where one can play with the spaces. But
now I believe that, in the end, these big spaces are
not satisfactory. I believe rather in the necessity for
the work of art to be presented almost isolated,
facing the viewer. That means returning somewhat
to the formulas of the era before the museums.

### Pontus Hulten

Most of the museums that have been built in the
last few years have been rather bad, but there have
also been a few very good ones. One that I would
like to mention is the new part of the Louisiana
Museum north of Copenhagen, which I think is an
excellent example of museum architecture: very
quiet, very efficient, beautiful light, natural light
and so on. And no ostentatious, flashy architecture. It's just beautiful and things look very good in
there and it's discreet and the nature around is
almost put to advantage by the buildings.

### María Corral

Museums are more and more taking the place of
civic centers, in a way, playing the role that
cathedrals used to play at a certain time, and
probably this is shown in the architecture of
museums. So Pei's building at the National Gallery
in Washington D.C., such a meeting point where
all the great events of the city are held, is much
more important than the exhibition spaces, is it
not? And we can find the same case outside the
United States, for instance, Stirling's museum in
Stuttgart. I am also worried about the way
museums are being built today, since I think that at
some point this whole situation is going to change,
or at least get more stable. And then all of these
people will find they have something absolutely

useless, a useless spectacle. I think that the battle with given spaces, with a determining architecture, for instance, the case of Reina Sofia in Madrid or the Castello de Rivoli in Turin, is always more dynamic and suggestive than the relationship that appears between the artists and a museum by Gae Aulenti or Stirling. I have no doubts about that.

### Johannes Cladders

I'm against the idea that contemporary art, for instance, is best placed in an old factory. The argument is always: This is a neutral place, a neutral space. I cannot see this. Very often the architecture becomes too strong. It overpowers the work of art which is in it, and that is a danger, in my opinion, because the architect is an artist too, I am sure. But his work of art has to function, unlike a picture or a sculpture which—of course—has to function too, but in quite a different way. They build their own work of art without thinking about the function of it. There is a danger, and the danger, of course, is: You can build a museum as a work of art, as architecture which is a work of art, but in the same way you don't recognize the necessities of art. You do not understand the necessities of paintings. In that case, the museum can be a very good work of art—the architecture can be a very good work of art—but it is not a museum.

### Richard Koshalek

I think what is important, and you hear this argument all the time about: We want an architecture that is totally neutral. We want an architect that gives us just sort of wide open spaces, and very comfortable spaces and so on. I am not of that opinion, actually. I would prefer to have problems with specific spaces within a museum and to have somebody, say James Stirling or Arata Isosaki, design it. I think the architecture should be a statement by the architect. I do not think it should be a totally neutral building, and I think that if it is a statement by the architect than we've got something to work with, something to rub against. There is some friction there that can be developed.

### Riva Castleman

Architecture is important. There is architecture that aspires to have its own identity so strongly that it overwhelms everything it is surrounding. Then there is architecture that is amenable to the purpose, to

such an extent that it is almost self-effacing, and this
is what I am talking about.

### Thomas Messer
My opinion is that the greatest gift that an architect
can make you is to provide you with a great,
distinguished space that itself would be a work of
art, and with which the objects that you present
would enter into an intense and meaningful
relationship.

### Jean-Louis Froment
I keep aloof from architectural problems. For me, a
museum is first of all what happens inside. It is the
artwork. It is the strength of the artwork. And
whether the setting is really magnificent, subtle,
elegant or vulgar, every conceivable adjective, made
by the greatest architects or extremely modest ones,
or with no architects, I think that we ourselves must
work closely with the artist and the works of the
artist. That's the first thing. So as to keep as far
away as possible from the building.

## 3. audience

### Marc Schepps
I think a museum should be judged not only by its
immediate success with an exhibition, with one
activity or another, but also retrospectively. Which
means that ten or twenty or thirty years later we can
judge the importance, on the level of the culture's
future, of the part it played at a given historical
moment, by giving voice to certain new tendencies
or certain new phenomena. Therefore, a museum
cannot only be judged on the level of the present,
but the activity of a museum must also be judged at
the historical and, therefore, retrospective, level.

### Lola Bonora
Our policy is to inform and to inform in a broad way.
In fact our director has always talked about information at a hundred and eighty degrees, meaning
open information in order to create a knowledge-
able public. And in fact, after twenty-five years, we
can say that people systematically come to look at
exhibitions.

### Gary Garrels
Museums I think have a different responsibility to
the public, and how one defines public is a problem-

atic issue. And I think the situation in the United States and the situation in Europe are very different.

### Evelyn Weiss

There have been a lot of changes in museums. I can speak more or less for Germany because it's the art scene I know best. And I remember when I started in the late sixties, the museum was quite a closed institution, and the museums in America were the great examples, and we thought we could never match them. But somehow during the development of art in the late sixties and early seventies—the period when pop art really came to Germany and Europe and was presented and so on—there was a real opening from the museum to the public, and the public came into the museum in another mood than before. And there was really a lot of life, and we had quite new problems to solve and to deal with. And then we had some examples from the States that we could adopt, and some things we had to do ourselves because our institutions are very different, basically different because we are supported by the city, the community or the state—the different "Lander"—and we are not private, which is a big, big difference in the structure of a museum. We had to solve our problems in another way. But the main goal was to open the museum to the public, so that the public could really think that these were their collections, their museums, their homes, that they could feel at ease and go around and look at the works and do what they wanted.

### Dominique Bozo

For me, museums are a permanent reference for the twentieth century by their quality, but they should be sustained by information that may not yet have been completely discovered. In short, a kind of pedagogy—to use a word I don't like—an audience that can be attracted by works of art and may ask for more things and information.

### Marcia Tucker

What happens, especially in the United States, is that if you are a curator and this is what you are most interested in and you spend years, as I did, doing exhibitions, working with artists, looking at work, the more you work, the older you get, the more adventurous you get. It's not the other way round. The people who are very young are much more self-conscious about what they do and try very hard to do things right. The older you get, the less concerned you are with any of those things, and the more you can really take chances in your work. As a curator, the more you take chances in your work the more likely you are to come into serious conflict with the institution as an institution, because if your shows are really difficult, really provocative, really controversial, if your shows are not funded and if they do not bring in big audiences, and if the institution has financial difficulties—which all museums do—then sooner or later your needs and the needs of the museum are going to clash. It is at that point that curators end up leaving their institutions and becoming directors. The reason they do not become curators at other places is because the same problem will arise. So you feel: Well, I will become the director of a place, or I will try to become the director of a place, and then these kinds of problems will not happen. And it is true, they do not happen, except that you are not doing the same work anymore. What you end up doing is raising the money so that other curators that you hire can do the work that you most wanted to do.

### Harald Szeeman

Of course today the museums have also became very avid to have a lot of visitors and all this kind of stuff. And for me, this is not so important. I prefer one guy understanding art than a hundred thousand paid tickets. But it becomes more and more difficult to find the money for this kind of thing.

### Diane Shamash

Publicity—I mean hype or marketing, whatever you call it—I think has really affected the art world in a very profound way. It has affected artists in their studios. It has affected galleries. It has affected museums. It has affected the whole network of organizations. It's very much like the fashion world. I think it is a changing phenomenon: something comes into vogue and something goes out. "Isms" —you know, neo-expressionism—or specific areas of artmaking are labelled almost just in order to promote a particular product. I think it is maybe a natural, if unpleasant, extension of our society. We're living in a very advanced capitalistic society where the value put on things is the value of the dollar. And I see the art world not being immune from that whole phenomena.

**Jean-Hubert Martin**
The mass phenomenon today is inevitable and, as you say, it is connected with getting grants, meaning that we cannot obtain the large sums of money needed today to hold exhibitions unless we attract the public, and this happens anyway since there's a growing interest on the part of the public for these exhibitions held in modern art museums. But I think that this interest from a larger public today for modern and contemporary art also makes it possible to create and promote other activities, to open small art centers, small museums that attract fewer people and make it possible to see works of art in a much more pleasant context with less crowds: that kind of more or less private situation where you can have direct contact and the chance to meditate in front of a work.

**Yves Gevaert**
When the public shows curiosity, I am all for it. And I devote all my time to it, when the public is really interested. But this public cannot be ten people at a time. But if at my place—well, I organize exhibits at my home—I organized a small exhibition of all the books and all the writings of Rodney Graham, and it was by appointment only. There were people who came here to the house, and sometimes I would spend hours trying to explain to them and have a real dialogue, and I think it is the only solution and the only way to reach an understanding of art. When it comes to the way the public is treated these days, I think it is badly treated.

# Chapter 5  the docents

**do-cent**
- that teaches or instructs

- - a person who conducts guided groups through
  a museum or art gallery and discusses and
  comments on the exhibitions

### Barbara Stewart

It is Greek and it means *docere*, and it means to lead or guide. And you draw out. In other words, I guess maybe it is the same thing you have been talking about: you draw information out of the person looking so they can enjoy it more.

### Claudia Wishnow

Docents connect the public to what is going on in museums. They try to guide them into some kind of an experience that is more than just "quickly looking." It helps to digest what they have seen, to put it into some kind of proper perspective.

### Kathleen Byrnes

Most of us have some kind of an interest in art, and that is why we got into it. I collect art, and I also buy and sell art on a small scale. I am so enthusiastic about art, and I wanted to do something for my community, and I decided I would do it in the area I was enthusiastic about. So my role here is to have people come in and look at art—especially contemporary art, which they so often do not understand—and make that connection for them. So they do not walk out of here saying: Is this really art? What is all this about? Because people are sometimes threatened or confused by contemporary art, and I want to make that connection so they will go out enthusiastic too. That is why I am in it.

### Kira Perov

*What is your interest in being a docent in this program?*

### Mary Drobny

To be able to share what I have learned with others. To get people in the community interested in our museum, to get more involvement with the community. I have an interest in art history, and I am involved in the City College art department and working there, and I realized that everybody in Long Beach is isolated. City College Art is isolated from California State University and from the museum and I'm trying to diversify myself and be involved in all aspects of art history.

### Ellen Breitman

The docents are the mouthpiece of the museum. And especially with contemporary art it is difficult for many people to understand what they are looking at. The docents have gone through a rigorous training program. They learn about art history and about materials and techniques, and for each exhibition they learn about the artists and what they are trying to do. And they can explain in general terms what the art means and why it is in the museum. And many times they have to deal with a very difficult problem of explaining art that is not easily understood to an audience that is sometimes even hostile.

**Russell Moore**

They are doing a most important service and at the same time…The museum staff also sees the docents as a professional part of this museum. They are the first to greet the public, they interpret the shows for that public, and they are our only contact with people who walk through the door.

**Barbara Stewart**

I feel that the docent isn't there to tell the viewer what is in the piece or to tell the viewer what the artist meant. We are not really there to interpret, but to help the viewer see what is actually in the piece and to get out of it what they can, depending on where they are in their appreciation of art and their open-minded view of their world. And depending on where they are coming from, I try to direct viewers to look at the piece itself and to draw from it what they can, ask questions. I like a dialogue with the viewer, rather than a lecture.

**Kathy Delap**

The docents take additional art history classes whenever possible, and part of the docent training is a section on art history, as well as some lectures on public speaking: just how to be comfortable talking about works that are in front of you. And then they do workshops on handling materials, on talking about art. Basically, with groups of people who are not familiar with looking at contemporary art, I see my role as trying to make them comfort-able. Very often people are intimidated by contem-porary art. They don't understand very often that there is a historical background to both the artist and his work or her work, and hopefully by sharing some of that information, they become comfortable enough about looking at it. And then we hope that eventually they will go on and read some material. We have material available in our bookstore, and we have sources of information on the kind of art that we are exhibiting, and we can suggest where they may go and find further information. So what you are hoping to do is to say to someone who comes into the museum that there are a lot of valid reasons for this being here, here are some of them. Whether you like it or not is not really the issue. The issue is for them to understand why it is being presented and hopefully to make them curious enough to go on to explore it further. Sometimes you do not know, of course, when you are talking to people, particularly with children, whether you have sparked an interest in this subject or not. But hopefully that is what you are doing.

### Ellen Breitman

Docents understand and try to tell the audience that each piece of art is open to interpretation, and they present as objective a view as possible. But some of their own feelings will definitely come into what they are saying and they are usually pretty open about that. For example they might say: I personally like this one for this reason. Or, I feel that this relates to the early work in this artist, although perhaps it's not visible to you. Or something like that.

### Barbara Stewart

We are each going to deal with it differently, depending on our own frame of reference and what we are comfortable with. I do not really like to get involved in saying this is what the artist meant or that is what the artist meant. I think that is very thin ice, and only the artist knows what he really wanted to do when he did it. I like to help the viewers look at what is there and find out what they think the artist meant or what the piece means to them. Perhaps that is the intent of the artist himself, not necessarily to say this is right or that is right or this is good, but to get the viewers thinking a little bit.

### Kathleen Byrnes

I think art is highly accessible, and that is how I see my role. When people walk in here, it is to make it accessible to them. I think art is very accessible, and I think most people have some form of art in their home, and they are not even aware of it. You know people put pretty things on the wall, whether they are posters or whatever. And just like music is readily accessible to everybody—there is no education required to listen to music and to enjoy it. You

might not know the composer, but you still enjoy it.
And I think art is like that. The color involved in
contemporary art especially. The color is so abstract
that it is open to all interpretations. So I think it is
highly accessible, but I think most people don't think
it is.

### Kathy Delap

But that is only really on a kind of emotional level
that it is accessible. I mean, to get beyond just one's
response to it requires input by the person. I mean,
you can look at—we have some pieces of outdoor
sculpture—we can look at what we would maybe
call "minimalist art" or something, and maybe we
would get an emotional response to it, or like it or
dislike it for a lot of reasons. But to really, I think,
understand what the artist is doing and why art is
important in terms of our time, then it requires more
out of the viewer. They simply have to know more
about it. You cannot make a critical decision about a
contemporary work unless you understand what the
artist is trying to do. And you do not understand
that—I do not think—totally just from looking.

### Barbara Stewart

Every work is influenced by the time in which it
happened and what was going on in the world. The
artist is responding in some way, just as the viewer is
responding. And what the artist has used to
communicate his idea or her idea is what I want the
viewer to recognize or focus on. Everything is
influenced by what is around it and how it is
presented. There is no way not to be a little bit
subjective. No matter how hard you try, you are
going to be subjective. It is impossible to communicate if the very words you use have some connotation, as well as the actual literal meaning of what
you say.

### Jan Lahey

If you can say: yes, I am seeing things from a
different point of view now, or I'm thinking a little
more deeply on this subject or that—the artist has
succeeded. And try and let go of your, perhaps,
preconceived notions that art has to be beautiful or
has to be something you would want to hang over
your sofa. There is lots more to art, and content may
be more meaningful to you than that beautiful
decorative piece that you also enjoy.

chapter 6  the critics

**crit-ic**
- one who expresses a reasoned opinion on any matter (as a work of art or a course of conduct) involving a judgement of its value, truth or righteousness, an appreciation of its beauty or technique, or an interpretation

## 1. role

### Benjamin Buchloch

I am a critic in the traditional sense, and I am working also as a historian, hopefully more so in the future, and I also work as a teacher. And I do perceive my work primarily as critical writing, critical teaching and reading and rereading of art history against the official versions that we have been acquainted with. I consider that to be a crucial task for a critic and for a historian, to rewrite the official version of art history, whether it is that of the immediate present or whether it is that of the past. I do consider the task of the critic to brush contemporary official art reception "against the grain," to quote a famous phrase. I think there is an essentially falsifying process going on in the reception of contemporary art that, I think, it is one of the tasks of the critics to oppose. The institution-alization of aesthetic practice is to be clarified, to be kept open, to be made transparent as much as possible by the critical writing. These are some of the tasks that I also want to perform in my interaction with students, and also I do consider it important—as far as the teaching is concerned—to open up avenues of historical understanding, to discuss and introduce marginal, so-called marginal, material into the art historical context that might be otherwise neglected or ignored altogether.

### Bernard Marcadé

And I try to stick as close as I can to artistic procedures. So I learn a lot from artists. A critic is often someone who teaches artists. Which means, he tells them: You are unaware. Basically, the traditional discourse of criticism is: The artist is unaware of what is going on in his art. And, basically, the critic is supposed to be there to tell the truth about what the artist is supposed to be unaware of. I claim that, on the contrary, I learn from artists, which means that just like a philoso-pher, like a scientist, the artist is someone who teaches me and who has some independent procedures, and so I'm very attentive, very attentive to art as a place of thought.

### Lucy Lippard

As a writer primarily because that is my medium. I intensely dislike the word "critic" because it puts you in an antagonistic position to artists, which I have never been. I have learned everything that I know about art from artists. I have lived with artists. I have organized with artists. And I see myself as a supporter rather then a critic, which is not to say a critical position isn't important. But, as you probably know, I am not famous for my critical positions: I write primarily about what I like, and I write about artists who are criticizing society. So artists are my vehicle for criticism, rather than my direct criticism of anybody's specific medium. So I see myself as an advocate basically: an advocate and an activist and a writer.

**Tommaso Trini**
I am more in favor of a creative relationship with the work of the artist. I am more of a writer, and I do not have a semiological or sociological structure as an objective basis for the reading of the work. I consider the artist the true historian. The only one who learns from the history of the past, who makes people understand what really happened to artists who came before, is the artist who works at a certain time in history and produces new works on the basis of what has been done before and also in perspective, with a vision of what should be the present.

**Donald Kuspit**
The title of my last book is *The Critic Is Artist*. I define myself as an artist. By which I mean that the title of my book is derived from the title of Oscar Wilde's famous dialogue—two part dialogue— called "The Critic Is Artist," and that had struck me always as an essential idea. An idea which is not undermined by its own ironical "tongue and cheek" approach. And essentially I take very seriously the idea that Wilde presents that the "critic"—putting that term in quotes, to keep it open ended—that the critic establishes, tries to establish the work in relation to the age, whatever that finally may mean (that is a changing concep- tion as well). And I regard myself as an analyst of art in relation to its age and as something of interest in itself, inherently of interest itself, although I have mixed feelings about that, increasingly. And also as a kind of therapist of art. I know that is an extravagant term, but I have a theory about art—modern art—being a kind of articulation of disequilibrium, almost the impossibil- ity of establishing unity or equilibrium, and the critic through the elusive, even devious, process of analysis—using a very wide variety of kinds of instrumentation—as in some sense, someone who

**61**

is "completing the work," unifying it, establishing an equilibrium.

### Dore Ashton

I have done all those things, and I don't have a role. I see my passion is to understand what attracts me, what moves me. So I guess you would say I am either an amateur or a dilettante.

### Bernard Lamarche-Vadel

I do not know if it is all right. I do not know if it is all right, but what I observe is that actually my work takes different directions depending on the moment. At certain times, it is true, I am more of a historian. Last year, rather two years ago, I published a book on Michelangelo. It is obviously the work of an art historian. I have just finished a book on Giacometti, which is a typical book by an art historian. But I am also very involved in the promotion of young artists, and that can be said to be specifically the activity of an art critic. It is true that the texts I write are, it seems to me, very much written texts. They are texts that aim at a certain stylistic quality, and so I think of myself as a writer who asserts himself as such, even when I write on art. I also think I have a certain talent as an agitator in favor of the artists I support, whom I defend, and whom I want to promote. Besides, I cannot conceive living otherwise than surrounded by works of art, which led me to put together a collection, a small one, but still a collection. Unlike many of my colleagues, especially in France, I do not disdain commercial activities either, which means that I think that my activity as a critic ought also to be remunerated. When it cannot be remunerated by the newspapers or the various outlets in which I publish, then I find it perfectly normal to do some trading, to buy and resell works of art. Which, by the way, was what my predecessors did rather secretly. Everybody knows that Andre Breton was a painting dealer, that Paul Eluard had been a painting dealer, that Apollinaire was a painting dealer. I think there, too, we have a French tradition.

### María Luisa Borrás

Art must be studied in its historical framework, of course, and therefore in its social, political and ideological context. Which means that art must not be considered an individual topic that is considered a science apart, but a subject that is tangled and necessarily related to the specific political and social situation of any given moment.

## Pierre Restany

By my training I am a writer, and it was certainly my destination to be a sociologist. So it is between these two extreme poles that the substance of my activity as a critic takes place, as well as its justification and motivations. For me art criticism is not a profession, but simply a way of life. And this way of life—and I am going back to what I was telling you at the beginning of this conversation—I ended up by shaping this way of life through a series of successive decisions. Not wanting to teach, not wanting to have a museum career: these refusals commit you to other things. So to live art as an existential dimension is obviously a general structural option.

## Guy Brett

A mixture of everything. I mean, I also like to try to organize exhibitions, as well as to write, especially if there is some work which I think should be seen and is not being seen. Then I will try to find a way of making an exhibition. I will take the idea around to different galleries and see if I can find a venue to show it, and I work independently like that. I have done that quite a few times. Not only, let us say, the work of intellectual professional artists, but I have also organized a number of shows of what you might call popular art, or expressions by people who are not artists but seem to me to be very significant.

## Jeanne Randolph

I define myself as using the medium of writing. The place from which I'm most allowed to speak is the publications which are usually open to critics. But I would say my relationship or my theoretical position would be actually that of a fellow cultural producer. And in a society—North American society—which I consider an embattled space for cultural producers to do their work as distinct from technological work, I'm one of the embattled people. I'm with those workers.

## Achille Bonito-Oliva

The art critic, as far as I live this role, must be a writer, a theoretician, an organizer. I believe that a true art critic must be a modern intellectual figure, because if it is not possible to make a revolution without organization, I also believe it is impossible to create culture without organization. But organization is also not possible without a theoretical project. Therefore the critic is a figure who must have all these roles within himself.

### Craig Owens
I do not think of myself as simply a writer, and then I do these other activities on the side, or as sources of income, but I feel that I am operating in this context on multiple fronts at once. And that is very important to me. So that I do not think of myself simply as a critic or as bound into that role, because, although my primary concern is criticism, I think that it is necessary to intervene in various points and at various ways within this whole art system.

### Catherine Millet
I think I do not do exactly the same work when I personally write an article that I sign, that I fully take on myself. And at that moment, I do the work of an art critic—which I try to do to the best of my ability—the work of a historian of contemporary art. Well, that work is different from what I do as chief-editor of an information magazine called *Art Press*. I think these are two relatively distinct functions; I have to differentiate between them.

### Daniel Giralt-Miracle
We art critics are sort of a new media. We are a media as long as we are the ones that establish bridges of stimulation and concern between art and society (school, general audience, university, people in the world of culture…). But we are a little bit like the agents who provoke a short-circuit and ignite an artistic situation. Or we might also be the big projection screen, a kind of loudspeaker for an artistic movement that is being born and for which we act as theoreticians, as systematizers, defining its doctrine and spreading it through our colleagues.

### Christopher Knight
I am a critic, but my biases are, I think, fairly clear in two directions. One is that my academic training was as an art historian. My graduate degrees are in art history, early twentieth-century modernist art, and that is definitely one bias that I have critically. The other one is one that I have been considering

only recently, only within the last year or two, and
that is as a writer, because I do not think enough
attention is paid—as far as art criticism goes—to
the activity of writing. I see criticism as a contract
between a writer and a reader. I don't feel my
primary responsibility is to the artist that I am
writing about or to the museum or to a gallery or
to history or so on and so forth. My primary
responsibility is to myself as a writer and to the
reader.

## Nena Dimitrijevic

Because, as you know, there is no formal training
or universities which would make you an art critic
of contemporary art. So in a way, one has to
choose to exist, to work, between all the current
theories—whether they come from, say, philosophy, psychoanalysis, anthropology, semiology—and
to find one's position within these ideas, which
means choosing a certain ideological position and
then applying it to art, contemporary art or living
art, just out of necessity to try a theoretical
discourse for current art practice.

## Thomas Wulffen

First as a critic, of course, then second as a writer
and the third one may be as a sort of an activist, in
a way, because Berlin is a special case for contemporary art. There is not much contemporary art in
Berlin, so I try to force them to present more
contemporary art. So my critique is against the
institutions and sometimes it is very hard, and that
is in a way my role as an activist, to force contemporary art more in Berlin than it is.

## Yves Michaud

Today the range of possibilities for the definition or
non definition of the critic… Basically, I would say it
starts with the "publicizing" critic, the propagandist critic, who is almost a commercial agent. Well,
I mean the one who helps to build up reputations,
who can also organize exhibitions, who can be an
appraiser, who can be a publicity agent. At the
other extreme it would be the art historian. The
one who situates what is being done in a more
general perspective, let us say in the spectrum, in
the whole range of artistic productions. And then
between the two, there are all the shades. From,
let us say, the propagandist to the one who claims
to be the arbiter of taste, and then finally the

dogmatic judgment of quality and then the objective judgment of a work's place in history. There is a whole range of positions. For my part I have always been embarrassed by this range of possibilities, since I do not consider myself a writer, although there is still an obvious pleasure in textually writing, in the textual work. I do not consider myself a propagandist either, but nevertheless, I admit, when one writes about a young artist, it is in order to make him known. So even with the best intentions, there is an aspect of publicity, and I confess that the more I have learned about art and the more my experience has developed, the more I have tended to place myself on the side of the art historian, meaning: an interest in locating a work in a larger context.

## 2. interpretation

### Bernard Marcadé

I have done like everyone else, all the critics of my generation; of course, I have been introduced to the social sciences, to linguistics, to psychoanalysis, to sociology, to the philosophy of current language, to all these approaches, in fact. But I think that in art there is a specific discourse, a way of operating that is specific, and that is the one that interests me, so I am very open. I always try to be. It is a hard job to be always open to the artist's work and position.

### Donald Kuspit

I think I began my criticality from an art historical philosophical basis which was always dialectic in orientation, but there was an emphasis, say, on the objective pole in which I thought it would be possible to analyze the work, to reflect on it in a Hegelian sense, that is to approach it through the negation of analysis, but then to perform a negation of a negation—a Hegelian speculation— and recover a sense of the work's concreteness, untranslatability and integrity and "autonomy." I have moved away from that, which involved in linguistic terms an interest, say, in the semantics and syntactics of the work. I've moved away from that. I feel that in some sense it is easy, increasingly easy to do, particularly in the eighties where we are in an eclectic period for art and while there are all kinds of meldings and complexities. One can trace them down. That kind of analysis structure is increasingly conventional. Also, it strikes me as inappropriate because I became increasingly aware of myself—in the process of making that—not particularly as a solid subject that was some kind of center for the analysis, but I realized that in making the so-called objective approach, I assumed, I presupposed, a kind of direct reciprocity between myself and the art, and then I found that that didn't exist. I experienced it as not-existing, and I also became concerned… As I began looking at, as I continued the process of looking at, lots of art, which I do "professionally" on a rather regular basis, I began to ask: What good was this art for me? And in general, for the world? What was its point? What was its relevance? I mean, apart from this wonderful circle of people called the art world and apart from that equally larger wonderful circle of people called the interested public—including the buying public and the general, educated public—I did not understand…I became increasingly at a loss to understand exactly what, to borrow from Norbert Wiener's book, what the "human use" of art was. And I became increasingly concerned with that. You can call it a "moral existential concern," although those are heavy words, and I wouldn't want to stress them because they imply a certain expectation of what the moral and existential would be. I didn't have that expectation. So what happened is—and what I am still in the process of doing is—I am trying to understand the work now through my own response to it, which does not mean that I do not make the other kinds of analysis. But I have become increasingly concerned with what the linguists call "the pragmatics of art" or, following the ideas of someone like Jauss in Germany (who is not alone in this), "the aesthetics of reception." I became interested in myself as, to use some of the fashion-able language, an "unprivileged" reader of the work. Whereas before, I saw myself as a privileged reader without realizing it: as someone who could understand the objectivity of the work, how it existed objectively. And I found by giving up this objectivity of the work, I somehow came closer to the truth of it, in a dialectical sense, by understand-ing it not as a lived experience that I had, but how it "expressed my needs," what my intention towards it was, all of which was obscure, which I had to bring out. And this is part of the reason I have become increasingly interested in a psychoanalytic understanding of art and an effort to understand the work of art in terms of what in psychoanalysis is now called "object relations theory," which does not mean that that is the end of my understanding. But I first have to understand what I myself want from art, the kind of "transferences" and "resistances" I have in relation to art—to use the primary language—and its role in my own primary process of thinking, so to say, before or at least simultaneous with my so-called objective analysis of art.

### Dore Ashton

Temperamentally I am anarchistic. I do not belong to any "ism." I am, for instance, politically radical, but I was never a member of any radical party. I am intellectually an eclectic, but I have always been interested in, let us say, formal problems in the arts. So I do not really want to categorize myself. Moreover, I think it is probably useless for a person like me, namely an art critic, to borrow the academic disciplines, poststructuralism or any of the "isms," because I do not think that is what is essential in responding to works of art.

### Filiberto Menna

I believe that the critic must, somehow, go through a moment of objectivity, that is to say he must declare the coordinates of his reading. He must make explicit the method he employs. He must, somehow, objectify his impressions, his preferences, the love and the hate which are born out of his encounter with the work of art. Therefore I consider it indispensable for critical analysis to go through a "cold" phase, but I believe this to be only one moment of the interpretation. I believe, following precisely Freud's distinction, that the critic must follow up his interpretation with the construction of a discourse. The critic who is worthy of the name always carries within himself the attempt, the aspiration, the desire to offer an interpretation of the world, of life, and this takes place through his critical interpretation. This is where the critic becomes more subjective, in following his own thread, his own vision, which not only concerns art, but life in its entirety.

### Benjamin Buchloch

I happen to be rather traditional in my assumption that works of art do have a historical reality. I happen to be rather traditional in my assumption that works of art do have a material reality, a

social-political reality, that they're not free-floating
signifiers, but they do have a referent within
history, and I do consider it to be the crucial job of
the historian and the critic to reconstitute that
particular context. First of all. Which does not, of
course, allow us to forget that while we attempt to
reconstitute this original historical material reality
within which the work has once functioned, that
reading since then could be obliterated. Clearly the
reading that we develop within a work now is as
relevant, is as important, as the reading that a
work might have received in the nineteenth
century. So our current perception, reinterpretation,
redefinition of meaning is an additional, a different
but essential, aspect of meaning to the historical,
material, social reality of the work in its proper
context. And obviously what I am sketching out is a
standard dilemma with which all historians and
critics struggle to some degree or other, and I think
it is a dilemma that one has to keep together. I
cannot decide either to take this position or the
other position. I think one has to struggle within
the two roles all the time. I object against the
relegation of objects to the past as much as I reject
the notion of an object that can be constantly
recreated and redefined. Between those two
extremes, I think one has to struggle.

### Jeanne Randolph

All of my work has been devoted to what I call the
politics of interpretation. I don't see interpretation
as an act in which you discover something intrinsic
in an artwork, and I don't see interpretation from
an external point of view—whether it is subjective
or objective—imputing an interpretation to an
artwork. I have been more or less studying or
making a theoretical position about the interaction
between the viewer and the artwork, and it is
possible to analyze that from a psychoanalytic
theory point of view, to point out the limitations of
theory, but to talk about it as an interactive
process.

### Pierre Restany

You are raising here indirectly a problem that is very
important: the problem of the boundaries and the
distinctions, the differences, between the art
historian and the art critic. Doing the historian's job
means above all trying to define objective criteria in
the fundamental presence of the work of art. And

this obviously implies a certain way of looking, a look that involves temporal dimensions and judgmental dimensions compensated by complex parameters creating the necessary corrections. Doing the critic's job means, I believe, taking risks and committing oneself, precisely at that moment. A work created at a particular moment is subject to that moment. It illustrates it through the whole possible range of metaphor: accurate or inaccurate metaphor, paraphrase, antiphrasis, allegory, mythology—there is an ample vocabulary to signify this kind of metaphorical relation with the reality of the moment. And this is where one finds the maximal existential impact of the work of art. And as long as it is perceived by its contemporaries through the dimension of existential impact, it is perceived in another way. I think that the art critic, when he adopts this subjective approach, only sanctions the existential value of art in the making. For me this is where the most thrilling moment in the approach to art, in fact, resides, also the one where the risks of error are the greatest. But then, you will allow me to claim for the art critic the right to be wrong, because this right to be wrong is also the right to life, and it is precisely the right to enjoy life, the pleasure of life.

### Catherine Millet

I would conceive my work as an art critic, also a little as a historian of contemporary art. That means that when I write on a subject, I try, at least as far as my sources of information allow, to be as objective as possible and as broad as possible. I think I absorb an enormous amount of information on the subject on which I must write before I start writing. Afterwards, once I start writing my text and entering more personally into my subject, I necessarily start forming my opinion. At that moment, elements that are my own must come into it, elements relating to my own subjectivity, to my personal tastes. But that only emerges after having been filtered in some way. I try to use exterior elements, objective elements, as much as I can.

### Lorenzo Mango

The most appropriate term is one used by Susan Sontag: the term is "tendentious interpretation." This means that interpretation cannot be free, since it is not true that one can say anything about anything one sees. At the same time, it is not possible to give a scientific, exact interpretation of what a work of art

is, either. What one can do is to analyze according
to the methodological tools one has, but also
according to the true roots of language, in order to
uncover its structures, right? But at the same time to
detour them. One cannot say that things are a
certain way simply because that is the way they are,
which is never true, but one cannot make up a
construction that totally falls outside the work
either. So one must build a construction, a theoreti-
cal-poetical construction which starts by strictly
adhering to the work, but must be, at the same
time, also very distant from it.

**Peter Frank**
It is a free interpretation. I don't go to the work with
an *a priori* context for interpretation or method for
deriving interpretation, although I am aware of the
various methods and modes of critical structure. I
would like to think that I use those various modes as
they apply to the work itself.

## 2. anthology

### Yves Michaud
I think that normally the critic must evolve. So he
may evolve in two ways: he may evolve first because
the world of art evolves and that gives birth to a
type of critic on whom I look ironically, but I do not
completely put them…That is what I call "surfer"
critics. And by "surfer" critics I mean those who ride
the wave each time. There are also "surfer"
galleries, which means people who ride a wave and
then, you know, when it is over they come back and
wait for the next one and hop! they are standing on
the next wave. And there are people like that in
France and everywhere else. In every country there
are critics who rode the wave of the seventies, the
wave of the eighties, the wave of the nineties. It is
an attitude that is understandable, and that may
even reveal a certain intelligence, but that intelligence may also be opportunism. And another way
for the critic to evolve is for his experience, and his
experience of the world, and the formation of his
judgment to evolve as a result of experience.

### Dore Ashton
First of all, criticism after all is writing, and I really
think that the only worthwhile critics are people
who write well. So if they write well, presumably
there is some personal style that evolves, and over a

period of time you can read the temperament or the personality of the writer through a thorough reading. Therefore, yes—there is a coherence, I presume.

**Pierre Restany**
The art critic has value only to the degree that his way of life contains enough elements, enough testimonies. I would not say exemplary elements, but enough testimonies to make it interesting as a parameter of judgement for others. And it is in this way that a retrospective glance over an art critic's work may be interesting.

**Craig Owens**
I think that I have not been particularly concerned with pursuing a constant line. However, when I do go back and look at texts, especially when they are to be reprinted, I am astonished to find a trajectory in my own career that I didn't really know was there. I am astonished to see a logic that proceeds from one text, say, to the next. Often certain problems will be introduced at the conclusion of the text—which is, of course, where you're not supposed to introduce them, but I always try to— and then I will discover that a text written a year later or whenever will pick them up and continue them. And I think that this is not always the case. I think that there are many ways of functioning as a critic, but in relation to this, I think many critics simply take their cues always from outside. They will go to see an exhibition here, and they will decide they have something to say about that or maybe they don't even have anything to say about it, but they want to write about that, and they'll do that. And then the next thing they will do will be over here. They will be bouncing around, and as a result their text reads more as a chronicle of external developments within the art world. Now I hope mine do too. I hope that because they are involved with engagement, with certain shifts within production over the last seven or eight years, that that will be legible from those texts. But I also have—and I'm often criticized for this, but it does not bother me—certain specific interests. In fact, they are becoming more specified for myself. So that I feel more that I am now writing some- thing that has a certain kind of continuity, and that the texts are episodes in something longer and larger. And they deal with specific aspects. I am

more aware now of their interconnectedness and what kind of logic holds them together. I think it is very important to do that. I think it is important: I think that comes out of certain theoretical and methodological concerns.

### Catherine Strasser

Coherence just means that a single person has written two hundred, four hundred, or fifty texts. Coherence is primarily that. I do not think you can work by saying: I am going to be coherent, and if I take from my entire body of work some other subgroup, then the whole will not make a sense any longer. Which means you cannot work seriously if you are obsessed with coherence. Coherence is a completely foolish question, because it is a question after the event, which comes only with historical perspective.

### Achille Bonito-Oliva

I think that this problem does not exist for criticism, or at least it is a question of a much subtler, more invisible coherence. First of all, one must look at what the identity of the critic is. If the critic is a militant one, that is a critic who follows the evolution of research, then he can very well go from studying pop art to minimal art, do not you think? Because after all they are all phenomena that belong to one cultural tradition, connected to historical avant-gardes. And therefore, there is a cultural coherence, a coherence connected to the context of the culture with which the critic deals. But if the art critic, after studying pop art, goes on to study Guttuso (do you know this Italian painter?), a neo-realist painter, I believe that in this case there is no coherence. I do not think the critic should utilize the alibi of art history to feel like a scientist who has a neutral relationship with the objects that are under his study. I believe that the critic has a relationship of complicity with the work he chooses and the artist he selects. Therefore, there is a responsibility connected to the object of his analysis, and for this I believe we must ask the critic for an underlying, structural coherence, and at the same time we must leave him the freedom, the cultural opportunism, to go from one phenomenon to another.

### Victoria Combalía

And we must discern between people who change

in a very artificial and banal way, just to join a new fashion, and people who—this anyway should be considered artist by artist…Because you can realize that there are people who, even if they are using different mediums, they always keep a world and a set of problems of their own. So I do not find it fair to ask the critic to be too coherent since it is true that people's taste changes. Taste is another issue. It is normal that you enjoy certain things when you are eighteen years old, and others when you are forty. As I always say, it is typical of young people to greatly admire medieval art or at least, that was my personal experience, and I absolutely hated baroque art: I found it ostentatious, cunning. Now, instead, I love baroque and mannerism. And then there are personal taste matters that must be set apart from your opinion about art movements. For instance, there are artists whose work I can admire a lot but they just do not get to my heart, I mean, emotionally. And this seems very logical to me. One can think of many possibilities and get to different conclusions. One can also change one's mind, and in that case I believe one has to find solid reasons for it. As the saying goes, it can be a wise thing to change your mind, and this is true. So nobody should be that rigid in asking someone to defend the same thing for all his life. Really, the more you know about art and the more you have read and seen, in particular, the more things you like. But also depending on your mood, your age, even your memories, your experiences, you may like some things better than others.

**Filiberto Menna**
If we examine the work of an artist, or of a critic, in the long run we discover an internal coherence in it, if we are dealing with a real artist and a real critic, that is, and not with critics who just follow fashions. But the point is this, in my opinion, that coherence must not be sought after, must not be programmed, because in that case we would have (how shall I put it?) the building of a character, not the history of a figure.

**Benjamin Buchloch**
The coherence that I perceive that I have in my writings is really the selection of work, the artists with whom I have been involved at the various stages of my critical practice of the last ten years. That has been consistent and coherent. I think I

have systematically supported artists over the last ten years who define aesthetic practice in a very, very broad range of terms with very, very different approaches and yet, at the same time, differ from other artists very clearly and drastically. And I do not think I have developed the opportunistic incoherence that might be the hallmark of the truly involved critic, which means the truly involved critic supposedly is subjected to the shifting of times and the flowing fashions within the art world.

**Catherine Millet**
I mean it may be necessary to distinguish two things. I think coherence is necessary, because if there is a personality behind these texts, if there is a thought, it has to be perceptible. Yes. Good. So I am for coherence. At the same time, it does not mean either that one should get stuck like that for good in an image or that one is not allowed to rectify one's aim along the way. For example, in my personal experience, when I first started writing on art, I was very involved in what was called concep-tual art. So I went all out, even between quotation marks, for that, for "conceptual art." When I reread my texts now, a good fifteen years later, some of them look to me really very dogmatic, or look to me, yes, a little rigid in the presentation. And even rereading them bothers me. I say to myself: Really, I would no longer dare to write this way today. At the same time, I think I must still take responsibility for them, which means that if someone should want to reprint them, I think I would let them be published as they are, because they are part of the development that was necessary for me to get where I am today. And so I hope I am less dogmatic today than I was fifteen years ago. At the same time, it may have been necessary, in order to think the things that I think today, or to like the things that I like today, for me to go through that dogmatism because it may have been a necessary evil.

**Thomas Wulffen**
That sort of anthology…I do not think that is really necessary. Maybe for two or three exceptional critics that could be good, but most of that is unnecessary.

**Lucy Lippard**
For me the coherence has to do with how you see

what you are doing, and what are you trying to do, but that does not mean…Coherence does not mean it is going to be that consistent, really. I wrote an article a long time ago called "Consistency in Small Minds" where I decided, I think at a very early point in my career where I knew nothing, that I would not be consistent. Looking back over twenty some years of it, I realize that I have been consistent but that I do not necessarily respect consistency for consistency's sake, or change for change's sake in artists' work. And different artists are affected by different things. If an artist stays in her or his studio forever, consistency is probably more respected and has more integrity to it. If an artist's work leads her or him out of a studio and into different parts of the world, then I think changing is more to be respected, and I would identify with the second part of that.

chapter 7  the media

**medi-a   (pl. of medium)**
- a channel, method or system of communication, information or entertainment—whether magazine, newspaper, radio, television or public  platform

- • the material or technical means for artistic expression

# 1. function

### Roman Gubern

I think that from their birth at the end of the
nineteenth century, radio, cinema and electronic
media (when I say, media, I'm basically thinking of
the electronic media of our times rather than the
traditional media like the press) have been double-
sided. They are or can be generators of artistic
contents or forms. In other words, they can
innovate, from Man Ray's photographs to Fellini's
movies: that is artistic production. But on the other
hand, there is a meta-language, because the media
talk about art—especially through coverage, which
was traditionally done by the press and is nowadays taken up by TV and all media—with a more
informative than creative approach. In fact, there is
an old misunderstanding that needs to be resolved.
The Art World, with capital letters (although I am
against this category, as I will explain in a moment,
but anyway, we are still using these semantic
forms), art as a category, is usually understood as
the individual and unique product that is exhibited
in a charismatic way—with an aura all around, as
Walter Benjamin used to say—in the gallery. And
therefore this antagonism appears between an art
that is unique, aristocratic, elite, with an aura, and
the reproducible and multiple product like movies,
posters, video, lithographs. Of course, I'm against
this. I think that ever since Walter Benjamin wrote
his famous book on the artwork in the era of its
mechanical reproduction it cannot be sustained any
more. But actually there is always a latent antagonism still alive, still to be surmounted, that makes
artists look over their shoulder with contemptuous
authority on media technologies, and media
professionals look with a lack of confidence at
traditional artistic products.

### René Berger

The media do not transmit, they transform. And
they cannot help transforming. It is not possible.
Why? You have a newspaper, a daily, even if is still
prestigious, with a rather dusty prestige like *Le
Monde*, it is obvious that nowadays the arts do not
occupy a major place in it. The economy is much
more important, politics even more. Therefore, the
media, newspapers and radio, but let us first talk
about newspapers, must consider the event in
relation to how often it appears. So if it is a daily, it
cannot do anything important, whatever the value
given to such and such an event. That is why more
detailed articles can be found in monthly maga-

zines, for example, or weekly ones, or in literary journals. But the most important media today, when it comes to the size of their audience, are obviously radio and, even more, television. Now what can the radio do in relation to the visual arts? Because now we are not talking about music, we are talking mostly about visual arts. Because of its specific nature, it can inform people about events very quickly, but in order to dramatize, as it must do, the radio is going to increase the number of interviews and, if possible, if there should be a scandal during the exhibition, it will be able to exploit the opposition of contrasting opinions. So when I was talking of transformation in connection with the coverage of artistic events, we can already see a distortion starting to appear. Why? Because several artists—rightly or wrongly, I'm not judging them—understand that they must be present during the two days before the opening, at least on the day of the opening, and eventually one or two days after: at the moment when the concentration of the media will make it possible to stress the most important event, or let us say the most spectacular one. That is obviously true for television. That is why certain performances at *Documenta*, or anywhere else—I am thinking of Tinguely and there are many examples, Fluxus, and so on—have such an importance for the media, which are instruments of spectacle, which means they have to translate a meaning into dramatized images.

### Giancarlo Politi

An art magazine like *Flash Art* and the big magazines has an informative function only. People normally think that art magazines create artists or movements, but in reality we record movements and artists, because we always, inevitably arrive after them. We cannot create an artist, because the artist has to be there already. We record artists and situations already existing. The ability, the gift of a director or a publisher of a magazine, lies in capturing certain moments right at the outset.

### Judith Hoffberg

I think a great deal of communication is not only through the art format itself, but through the media that express and interpret it. And often, the "big four"—those large art magazines in the United States—and the others that represent established art neglect the device though which artists communicate. In other words, the eyes. And I think a lot of us forget that we have to see art in order to express what is important to us.

### Willy Bongard

I consider myself first of all a journalist. I do not like to be called an art critic, as a matter of fact. I prefer to be called an observer. I love to observe what is happening in the world of art. I am just curious, very curious about art, as well as about what makes art such an interesting subject, object I mean, as far as artworks are concerned. I am still very interested in the economics of the arts, but basically of course, it is the arts I am interested in. I think that it is kind of by necessity if you get interested in the art world, you end up either hating art or loving art. I mean it is a very mixed relationship. But again I am not a critic: I am an observer, a journalist who tries to find out what's going on, what is happening, and to follow the activities of artists as well as galleries, dealers, collectors, museums and critics. Critics are part of my observation.

### Tatsumi Shinoda

I define myself as editor-critic. And this is one of the reasons I am interested in the art world right now, because editing is quite interesting and important work in this situation where we can see that no further dominating tendency or movement is to be found. So to edit art history or Japanese culture or to edit this living world as a text is a very important job to me.

### Robert Atkins

I am trained as an art historian, so I think of myself as much as an educator as anything else. I teach as well as write, and I am really most interested in writing for non–art world publications. I would prefer to write for a magazine like *Time* or *Newsweek* say, rather than *Art in America* or *Artforum*. I think the best thing that I could do would be both to explain art to people who are little less conversant with it and to get them excited about it. I mean, as a teacher I feel like a real proselytizer.

### Mary Ann Staniszewski

Using my various tools as a kind of communicator

to deal with political issues is very much a part of my work. So it's, you know, the teaching and the writing, this means sort of manipulating the media or vehicles of communication to deal with issues that I think are very much connected to art, and it is also tied into the way I think about art. I do not think the conventional way of thinking about art and culture in this society, in the United States, has emphasized or recognized the way artistic and cultural communities have functioned historically. Artists and writers and intellectuals have always sort of banded together and worked for social change. They have created beautiful objects, but it has all been part of a kind of avant-gardist agenda. So I see all of my work within that sort of multifarious constellation, and none of it is perfect, and none of it is quite there yet, but it all provides different benefits for each institutional frame I am working within.

**Jef Cornelis**
For me personally, one cannot talk with a work of art. It does not talk. It remains silent. But I think the media are in a position…rather the media have no rapport with the works of art, it just does not work. Why? Because the media are kitsch. They can only create kitsch! But there is an instrument that can be used in order not to talk, but to talk with silence. I just mean that an instrument such as video or film gives you the possibility of seeing a work of art in a different way. Yes, then it can be isolated. But I am talking personally now. It is an opportunity that may be contrary to what we think, that it means to vulgarize, etc. I think the media, or rather television, has another possibility: to rediscover another rapport with the work of art. And that is something I have always looked for. We should not talk in theory but in practice. For example, if we talk about people who have done installations or experiences—like James Lee Byars, for example—if we examine this over a period of twenty years, all these installations, or rather these gestures, the games he was doing with the spectators…A film or video that could somehow record that gave the opportunity to take more of a distance. I do not know if you understand me. The camera (or rather, the media) offers the possibility to keep one's distance. And that is what I have always looked for, because I am someone who is always afraid of spontaneous thought. So the camera provides you with distance.

**83**

And for me, it was precisely an opportunity, because I am a voyeur, a real one, to keep my distance by means of that instrument. And I have always very much liked this kind of investigation, because it has given me the chance to work with other people on the same subject, which at a certain moment, recorded on film or magnetic tape, offers a chance to rework something that is closed, that one does not know, that one cannot introduce into a work of art. It is done! So it is closed. I do not know if I am making myself clear. An artist is not going to explain his work to you, so, does that mean he is not an artist? He shuts up, he wants us not to understand. Dialogue is ridiculous. So it is a paradox that the media want to explain and are unable to…to assume that paradox in itself will provide an opportunity for preserving better what was finally enclosed in the work itself.

**Geneviève Breerette**
I am well aware that when I do an article on a young artist, I am aware that it is a big help to him. So I am careful to stand up for values that I believe in, that I believe to be correct. If I take a young artist, for example, a very young one of the new generation, well, if I stand up for him, it is because I think there is something in him that justifies it. Now I may be wrong or I may know nothing about it, or we may only know in a hundred years, I do not know. Yes, of course, I have a responsibility, I know it is going to look good in his resumé. I know that later on he can show up with his whole portfolio and say: Look, so and so in such and such a newspaper had this to say about me—and so on. And it may help. I know it helps for sales. I know very well that when I do articles on galleries, the people go to see the shows. I am aware of all that. So what else can I tell you? Well, I have a commit-ment there. I make choices. I have to, because anyway it is impossible to report on everything. First of all, because it would be a pain in the ass. I could not do it. I make choices and I make these choices on the basis of my knowledge, of what I have acquired during twenty years of connivance with art.

**Brian Wallis**
I think one of the important manifestations of all these activities in the eighties was to muddy the

distinction between those categories that you just
mentioned, and to assure that it would no longer
be possible to neatly classify people as critics, as
artists, as editors, as curators. Seems to me that if
you consider the function of or the process of
doing any one of these activities, it is pretty much
the same. It is a matter of selection, critical
evaluation, analysis. So whether you are approach-
ing culture as an artist, as a critic, or as a curator—
basically the processes that you go through are
similar. And I think that for the artists and the
critics and the other editors that I worked with in
the eighties, it was the recognition that all those
activities were essentially involved in the same
project that was very important.

## 2. format

### Giancarlo Politi

There are two types of art magazines. The magazine with a very small structure, published by an
artist or a critic, or in any case by one person, and
this is the type of magazine of which we have—
perhaps a historical type, like, for example, *The
Minotaur* or classic magazines such as Man Ray's
magazine or Duchamp's magazines. After the war,
the art system naturally became more perfected
and more developed, and magazines with modern,
efficient structures were born. *Art in America*,
*Artforum*, *Flash Art* are the expression of this new
necessity to reach a vast public, to supply information which must be pretty serious and scientific,
but journalistic at the same time. Therefore, I
would not say that the way of making an art
magazine has changed, rather the landscape has
changed, it has grown. Until the Second World
War there were only magazines with a small
circulation, made by artists, made by just one
person. Since the war, especially since the sixties,
the art magazine has become a commercial
enterprise, as well as a cultural one.

### Willy Bongard

You have to differentiate between media and
media. As far as television is concerned, at least in
Germany, television is there for entertainment first
of all. And I don't think that there is a real impact
although many, or most, artists love to be on TV, as
do collectors and museum people. Art magazines
like *Artforum* or *Art in America* or *Flash Art* are

**85**

much more important than the true mass media. I mean they have a relatively small circulation. The *New York Times* is probably the most important medium in the art field in New York. Although I remember I asked John Canaday, who was once the most powerful art critic in the **New York Times**, about his influence and he said: "Well, I can pack a gallery by a review, but that is about it. I do not think I can really influence the course of art and create or really start a movement."

## Teri Wehn Damisch

I have always tried to dissolve the barriers between the hierarchies, and that is what my television programs have been about. I refuse the notion of high culture and low culture. Every time that a feeling of high culture or a museum has tried to impose something upon us in television, we have— on the contrary—tried to be iconoclastic. Sometimes with the result that the museums, or certain persons in the museums—certain stodgy, reactionary persons in the museum—tried to "get our goat."

## Jean-Pierre Van Tieghem

It looks to me like it is a problem of strategy for someone like me, who is not at all a creator, but who is in some way a transmitter of things through different means. It is obvious that when I write a text in a philosophical publication, I am going to get to the bottom of the problematics—philosophical, theoretical, ontological, phenomenological, whatever you like—to try to reach the basic problems in this discourse. Because I know that when I write in a philosophical journal, I am addressing a particular public, which is a targeted public and needs that kind of theoretical thought. When I write in a publication—let us take a review in France that has a good reputation, *Art Press*— when it comes to writing an article for *Art Press*, the thing is to try to get as close as possible to the artist's work. Not to translate it, but to get inside it and give a certain number of analytical connotations concerning that work. I say analytical and no longer theoretical. When I do a broadcast on radio or television, I cannot give analytical or theoretical connotations). In radio and television there are the big public networks, there are some more or less specialized networks, and even there the language changes. When I work on a very public network, then I sweep away anything theoretical. I sweep away anything analytical and try to emphasize the relationship between the work that I see in an exhibition—in a museum, in a Kunsthalle, etc.—and the person who has done that work. So it is a relationship, a very immediate relationship, to reveal through somebody's discourse something that is there, that is there once and for all in the form of pictures, in the form of sculptures, etc. And on other radio or television networks that are a little more specialized, I allow myself to go further. I may allow myself to hold a more perverse discourse, which may try to reveal a little of the underside. The underside is interesting too. But the underside is not the voyeur side; on the contrary, it means giving additional elements.

### Josep Iglesias del Marquet
The way this information is treated grows or decreases in depth depending on the medium that carries it out. Magazines—which I know well—not only accept but demand a deeper and more elaborated treatment than newspapers. The newspaper reader is a person who has little time and reads fast, while it seems clear that the one who buys an art magazine has more time to think over its contents. And this calls for—or reminds me of—the need for a visual document. This is the basis and foundation of television. On the other hand, radio asks for words. In radio, too, it is impossible to grow deep, to reflect. Facts must be given very clearly and with the reasons that justify them, but again one cannot take too long because the information is very fast.

### Shelley Rice
I mean if I feel like a piece is being changed in meaning or the artist is not being treated fairly in the piece or something, I will just take the piece back and won't let it appear. But it varies. I mean, I have had very, very good relations with certain editors for close to a decade. I have had very bad relations with others. Basically the stuff of mine that appears tends generally to appear unedited. I do not really need a line edited here and there, a word changed, and people usually call me up and ask me. Like I have said I guess that there have been various situations, most often in the art magazines, the specialist magazines, where the editing was just more than I could handle.

**87**

### Paul Taylor

In between himself and the reader is the editor. So you have to jump through hoops to even get to an audience. Because I am at the moment writing an article about a rather esoteric subject for a popular magazine and the editors—who really do not know anything about art—are continually trying to tell me what to say. And so I am aware that I have got to get past them to spread my point of view out among the masses, as it were. They keep a pretty tight control over what can be said.

### Rosa Queralt

A very important aspect of radio is the way you present things, like in most media. But in radio, that way is more important than in the rest of the media, and I think there are artistic products that just cannot be treated through radio. For instance, I am thinking of the minimalist exhibition that was held in Barcelona. At that time, I became obsessed with the possibility of talking about it in my radio program, of explaining minimalism through radio. I was very stubborn, so I worked a lot and did all I could to explain minimalism, and finally, after the program, my colleagues told me: "How boring it was today!" And other times, you would go out for lunch with someone, drink a bottle of wine between the two of you, and feeling happy and not having prepared the program at all, the program would turn out great.

### David Antin

If you talk to one person, it's like writing in a different medium. That is, in other words, if I have a conversation with you or I have a conversation with somebody else, there might very well be significant differences in the frame of the conversation. So, a newspaper has very rigorous constraints. My sense is that…like in the *Village Voice*, when John Perrault wrote for the *Village Voice*, he was a very good writer within the *Village Voice* possibilities. John Perrault knew he did a very good job as a newspaper writer. Lawrence Alloway writing in *The Nation* did remarkably well for writing in *The Nation*, and I think that Kozloff did too. Writing in magazines like those, that is weekly magazines, it is very hard to deal with that, so just like the *Village Voice*…There are some kind of magazines in which I do not think it is possible to say anything intelligent at all, and this is probably true of *Time* and *Newsweek*. I do not think it is possible! *Time* and

*Newsweek* make it impossible to be intelligent except under very rare circumstances. Newspapers are obviously very tight on "occasionalism," and they make intellectual criticism impossible in the United States. American newspapers are very different from European newspapers. A French newspaper can have very intellectual criticism. So can a German one, or an Italian one. But you can't have intellectual criticism on an ambitious order in a newspaper in the United States. It is just not possible! And that is because of the editorial board's view of the intelligence of what they regard as their public.

**Roman Gubern**

The intellectual who is doing art criticism or the person with a background of humanistic studies—which I do not attack, I am very respectful towards it—must be aware that he or she has to invent new ways of communicating, new communication models of the writing forms, to be able to transmit new subjects, critical or otherwise, since criticism is basically a meta-language. Because as a critic you find yourself speaking of art pieces that are very often images (they might be music, but are often images), and to communicate you are using images as well. From that moment the speech becomes a meta-linguistic speech, because you are talking about images through images. This is a very interesting process, but I find a lot of incompetence among our theoreticians, writers and critics, because there is still mistrust and, most of all, great ignorance about the nature and effects of new media.

**Heidi Grundmann**

For a while, it was, for instance, very important for an artist to be written up in magazines, to have his or her own critic interpreting his or her work—also in catalogues and so on. But I think that that has become a little bit less important than it used to be. And I think in the daily papers, there is a kind of boredom now, with writing about the many exhibitions taking place. And I think one has to find other forms and more differentiated forms of writing about art in the different print media and make many more differentiations: whether it is for a daily paper, whether it is for a weekly paper, whether it is for a paper that is more interested in economic factors, or more in philosophical factors, and so on. I think that there have to be more and different forms of dealing with art, and that has not been achieved yet. There is a certain jargon still in all publications, which makes it very difficult for the audience, and especially the wider audience, to relate to work that basically would be very easy to relate to, because there is now a lot of work being done that is basically easy to relate to for an audience, at least at a certain level. So I think new forms have to be developed in the print media, and the same basically also goes for radio and television. Radio has, of course, the advantage that it can present a person. The voice of the person, the ideas of the person. It can give a lot of time to the voice and the ideas, and that conveys a lot. And television, of course, is much more expensive, so you do not have so much time. But again I think one has to develop other forms, because it is simply not true that because television can be received visually it conveys, for instance, painting or installations or sculpture very well. It does not convey these things very well, and one has to find new forms. And I think it will be very, very important that art institutions make a special emphasis on educating young art historians and so on in these different forms of dealing with art in the different media.

## 3. fashion

### François Pluchart
I think fashion is something that, first of all, is inevitable. It is part of people's desires, so we have to live with it. Fashion is precisely what allows thoughts to develop, to be transformed, which means that one kind of expression gives way to another one. Well, fashion is obviously completely amoral, which means that a fashion of very high quality may succeed a mediocre fashion and vice-versa. Because it is the fashion, people feel like joining in. Well, at the same time fashion is extremely healthy because it makes it possible for things to convey new ideas and it makes things move and thus allows new creators, whatever their form of expression may be, to find an audience. Because without fashion, that is, within a totally repetitive system, there would be no possibility of creation. Now I believe that, in fact, works of creation always come into fashion at one moment or another. So that is where the big issue is: At

what moment does a creation become fashionable? Does it become fashionable the moment it is
created, or will it become fashionable half a
century later? A few years ago, let us say eight or
ten, Marcel Duchamp was very much in fashion. He
was very much in fashion fifty years late. It certainly
would have been much better if the readymade
had been in fashion fifty years earlier.

### René Berger
When we speak of fashion, we imply a transitory
phenomenon in relation to something that could
be stable or has been stable. As for me, I think that
fashion has become the model, the fundamental
mode of existence. There is no longer anything
stable, and therefore fashions are no longer
opposed to stability, essence, substance. What we
should understand today, is that fashion has
probably become the most important existential
dimension, and that it is no longer opposed to the
idea of an essence that I believe to be obsolete.

## 4. network

### Heidi Grundmann
There is not just one network, that is, there is
always the one network of the international art
market, of the big names and the big important
exhibitions and so on. But I feel that there are more
and more networks developing and artists are part
of different networks. The same artist is part of
different networks, just as people now are also
making art that is placed in completely different
contexts by the same artists. The same artist moves
through all kinds of contexts and positions his or
her work in different contexts; he or she also
becomes part of different kinds of networks. But of
course, everywhere it is necessary to have networks
because an artist in isolation is now even more
unimaginable than it ever used to be.

### Paul Taylor
Artists should try to take more control of their
work by creating their own independent media.
What we saw with the punk era was many, many
bands forming independent labels, so they could
distribute their own records to radio stations and
disc jockeys and so forth, circumventing the big
dinosaurs, the big music companies. We also saw
that with publishing quite a lot ten years ago,

independent publishing of magazines, of which mine was one. It's quite difficult in this sort of corporate environment to act that way, but independent production is one way in which you can really do it. Otherwise you have got to make work that is good for being cut up and for being censored and see that somehow or other its message still survives in the brutal environment of the mass media.

**Mary Ann Staniszewski**
You can look at the various historical avant-garde movements, like De Stijl, surrealism, dadaism, the Bauhaus, the Soviet projects, all of these groups were involved with a transformation of culture on a much broader base, and so much of the legacy of that is to work with the variety of means that one has, that one has in terms of culture. It could be exhibitions, it could be using the media, it could be making design objects, it could be writing certain kinds of poetry. You would use the popular press, you would use the arcane presses, you would produce journals and periodicals, you would work with picture magazines. Artists worked with radio, films, and so on. And then there were activist demonstration types, parades in the street, street theater. All of these things are culture as I under-stand it, both high and low. And in terms of the way it has been received historically here, in terms of seeing history, and in terms of the way the actual system has been set up, it has been reduced to a picture market. Not that I do not think pictures are terrific or that paintings are terrific or that they are not important, I think they are very powerful and very important components of culture, but they are not the only components of culture. In the eighties we had this sort of grand magnification of that way of seeing. It is a sort of model, an exaggeration of all those sort of modernist myths about what art is: we lived through it in the last decade.

## 5. audience

**Shelley Rice**
Obviously the media have a certain amount of power in this culture. In fact they have immense power in this culture. It depends on where you write, what for, etc.. But, for instance, when you write for the *Village Voice*, you are reaching a very

immediate audience. And you can see very immediate effects. When you write for *Artforum*, it is much less immediate: I mean everyone in the art world then perceives you as powerful, for what reason I have never been quite clear—just because everyone's perceiving you that way. And it is a sort of an "in group" type of situation, and you have a certain amount of power within the "in group". It is not the same kind of direct social impact. The magazines and newspapers in America these days are terribly conservative. Everybody's got their formula, and nobody wants to break it, and I think that's really where a lot of problems are stemming from in all this. Because the public really gets off on this stuff. I used to get letters from Bronx house-wives who used to read my columns, and they loved the stuff, they thought it was fabulous. I wanted to…when I was with the *Village Voice*, one of the things that I wanted to do was branch out, because I wanted to deal with photography and culture. So I wanted it to be in a social criticism section. I wanted to be in a news section. I did not want it to be in the art section because that automatically sets up a context and a whole series of associations. When I went to the *Soho News*, I managed to get the column that I wanted, but it had to be in the art section. It was not moved over to the news section, which is where it should have been. It is general interest, a basic education and discussion about what is going on in the world. That is where I thought it should be, because I was writing about photography. And again this is also personal, because I tend to… like I said, I really love communicating with people, and unlike many of my colleagues, that is my primary aim: to commu-nicate with people.

**Josep Iglesias del Marquet**
You cannot confuse the reader. You must facilitate as much as possible the reader's understanding. You cannot assume that the general public has that basic knowledge that is taken for granted among specialists. That is basic. Therefore, it is necessary to link criticism, commentary and information to references that are known or can be recognized by the reader or by the listener. In other words: you cannot start from facts and axioms assuming that everyone knows them. This is simply not true. And you must do so for many reasons, for instance, not to frustrate the reader. When someone does not

understand at all something that s/he is seeing or reading, s/he tends to reject it, leave it aside: s/he does not want to deal with it. The goal, in this case, for any specialist—in economics, movies, music or, in our case, art—is to have a pedagogic approach. You must connect people, and their reality, to the issue that is being considered. That is why very elite criticisms interest only very few people. This is not the function of the mass media: they must interest everybody.

**Robert Atkins**
I feel extremely conscious of my audience in terms of not only responsibility, but I wouldn't be doing my job and for instance…I write for *Newsday*, it is a Long Island newspaper. It is a very large newspaper with a million plus circulation, owned by Times-Mirror. My work is edited very carefully, and if I make references to obscure Renaissance paintings in connection with some twentieth-century artist's work, my editor, and or any of the copy editors, is very quick to point out that I am making a wrong supposition about this audience: I am speaking above their heads. You know, I'm missing my "cues" here. I am not dealing in a responsible way with the audience. When I do write for this daily newspaper, then I feel like an art journalist and I can never forget for a moment who I am speaking to.

**Geneviève Breerette**
Here there is already a difference compared to a specialized review. In a specialized review, one is addressing a very restricted public. Now my newspaper prints over 400,000 copies, almost 500,000, it depends on the season. And so I must take into account a larger public, which means that each time there is a need to make a certain number of references that one does not have to make in the columns of specialized newspapers and journals. I remember the term "conceptual" for example, even the "minimalists." Not long ago there was a problem with the term "minimalist" and so it had to be explained. What "minimalist" means has to be explained every time I use it. So when you have to explain every time what minimalist was, what conceptual was, etc., it gets a little complicated. Well, that is one point.

**Michael Brenson**
I am much more comfortable writing for a general

audience. I think that when you write for a trade magazine the language does change, and I always felt more restricted. I think that the possibility of writing for just intelligent people—who might care about art or who might not know that much about it—is something I like a lot.

### Rosa Queralt

I think that consumer habits have a very direct influence on mass media. This is one of the reasons why the presence of the arts in the media is very scarce. When an ad arrives, the art column is the first thing to go, because the audience does not demand it.

### Jef Cornelis

I never succeeded in having an audience for what I was trying to…I wanted to find an audience through television or cinema—to find an audience, because it does not exist. Whoever tells you that is fabricating an illusion.

### Brian Wallis

If you are a writer or an artist or any kind of communicator, and you do not recognize the difference in the various audiences that you are addressing, then you are sort of bound to fail in certain ways. You're bound to have your work misread or misunderstood. And again, I think that that was part of the problem in this controversy over censorship. Works that were created for a very particular audience—Robert Mapplethorpe's photographs were in effect created for a particular audience that understood these works, that understood the codes and understood what these photographs were meant to mean—those works taken in another context by another audience meant something entirely different, and that is where the problem arose. So, similarly, if you are a producer, a writer or an artist, and you have to communicate as directly and persuasive as possible to an audience, I think it clearly behooves you to understand the interests of that audience, the constitution of that audience, and why they would be interested in listening to what you have to say.

chapter 8  epilogue

**ep'i-logue**

- a short poem, speech or dialogue addressed directly to the audience after the conclusion of a play

•• a concluding section serving to round out or complete the design of the work

**Joseph Beuys**
First we have to find out what is the dilemma
within the art and the artist, and then we can
slowly speak about how it is transferred-this
principle of arts, artists—and galleries and museums and critics and media are all a side effect.
Because most artists are not interested in strug-
gling, in coming through with another understanding of art which is related to every work, to
everybody's working place, to everybody's problem.
There is the theory developed within the artists and
within the art, also not within the art critics, even
not within the galleries—which are only caring for
a kind of transfer, let us say market processes and
so on . So I see there is no really organic development of art in our society.

**Daniel Buren**
What you call a network is certainly what we called
fifteen years ago, more or less, the "system." My
point of view has always been very direct about
these meanings. I always thought, and I still think,
that this huge network obviously includes the
artist; so that gives already part of the answer to
the question. I consider myself, if I am an artist (but
let us say not, to play with that), being an artist
working in this art world, part of this network. In
other words, I do not see the system, or the people
from the media, or from the museum, like enemies
or like people outside of what I can do. So that is a
basis, and that basis is very important politically
speaking, I think, because there has been a
tendency in the late sixties, early seventies—due of
course mainly to artists—to show themselves as a
group outside of this system. And I was always in
conflict even with artists about this point of view,
because I think it is too easy to see it like that. So I
think this network finally is part of the work
everyone is doing. In other words: from that point
of view, I think it is better to have a critical point of
view about this network than to think that this
network exists but is like a detail, and then you
might be completely trapped by that network,
which I think unfortunately is what happens most
of the time. So, to say it in a different way, I think
the relation of all these elements is very important
to know. I think that everyone should know
absolutely as well as possible, how all these things
work. And when I said should know, that also
means you have to always take care, because
things change. Nothing of this…not one of these

parameters remains the same all the time. So you always have to keep very closely in accordance with these movements, and then I think a work can be done with more consciousness. Because, if you do not do that, of course works of art are made, but they are eventually completely…By force of being ignorant of this network, they might be completely destroyed or at least completely diverted. So I see all these things, and even if you want to make a criticism, I think you have to know even better. I think and I see all these elements absolutely tied to each other, and we have to work with them. So as an artist I cannot ignore these parameters, and even if you want to be detached, you have to know even better how they function.

## Luciano Fabro

The problems of the rest of the world do not apply to Italy, and especially to our generation—to the artists I was talking about at the beginning—because when we started working we did not have a gallery. In twenty years, thirty years almost—well, twenty-five years—of work, we did not have a gallery. (And you can see it even now: these artists are still free floating.) Both because of the work we did, which was new and did not interest galleries, and because there are no museums of contemporary art in Italy. And because collecting started very slowly in our case, we have had to invent on each occasion a way of dealing with our work.

## John Baldessari

I think a "mysterious consensus" can operate in a village-type of situation like New York, where everybody…It is a large village of course, but everybody is very close together.  So things operate a lot by what we call "street talk," one person hearing another person, another person, another person. I think that is the "mysterious consensus" that Castelli is talking about (cf. Chapters 1 and 3) . Writers need people to write about, dealers need artists to sell, collectors need art to buy: so there is always a market, but it has to be a market that is always fueled. You have always got to have a new model car, and so on, or hot stock, or what have you. That is exactly what makes the stock market go up or down. Talk. Optimism or pessimism.

**Regina Silveira**
The problem begins when some roles take over other roles. For example, when the market infiltrates into the museum ideology or into their operating system and programming, or when art critics follow the needs of the art market. I think that what happens when one of the roles takes greater importance and expands is that it can interfere and devour the other roles. I think that in the eighties the market had an influence even on cultural institutions like museums; many museums scheduled their exhibitions based on the leads of the art market, and especially in relation to the need for establishing new values.

**Fernando de Fillipi**
Let us say that there is a kind of reproduction of the "political" inside the system of art. That is to say that the politician, because of laziness and because he has given art a marginal function, a minimal function, is not himself concerned with it, and he delegates to those who, either out of opportunism or because they somehow want to,, enter the party determined to organize events. Now, for example, there are parties which have achieved a lot of prestige in the political sphere, they are everywhere, in every local administration, in the government, and therefore they need people to take care of this sector. Many of them are not experts, they are totally unprepared, and they delegate, in a way, to the so-called specialists. These specialists utilize public money without having to account for it, they are not required to— there are exhibitions which have cost a million dollars, a million and a half, totally useless, stupid exhibitions, but nobody is accountable for them. It is not like a museum which has a certain annual budget. These are local administrations, city structures, which have this money and spend it in this way. Also there has recently been a big phenomenon, the so-called "Nicolini" phenom-enon. Nicolini was the cultural commissioner in Rome and, through his cultural activity, he went on to have a big personal success later in political elections. He created events and made people go out into the squares, he created the so-called "culture of the ephemeral," and a series of situations which certainly got a big response on the spectacular level, but were not of a high quality. This triggered a series of mechanisms thanks to which the office of the head of the local cultural

commission is now very sought after, but naturally all these people in office have no understanding of culture. For example, the president of the Biennale changes every four years, the last president before Portoghesi was a historian, an expert in the history of Italy. He knew the history of Italy extremely well. He would have been a perfect director of a "Risorgimento" museum, but since he was a republican, and they needed a republican as the president of the Biennale, they elected him: his name is Galasso. He is a man of genius in his field and a very capable man, but he had to start from scratch, to understand what art history was, because he did not know anything about it. To make a long story short, we favor the politician over the professional.

**Adrian Piper**
The history of western art, particularly painting but also clearly sculpture, is in my mind the history of engagement with political and social content: it is in Géricault and Delacroix and Goya and Picasso and lots and lots of others. My feeling is that that is the tradition, that is part of the idiom of art and it always has been. In my mind, a very interesting permutation took place when this country first became aware of modernism, and, for me, the key moment was in the fifties with McCarthyism and the simultaneous ideological move on the part of someone like Clement Greenberg to define modernism and to define formalism in such a way that political content was not only unimportant, it was actively to be discouraged. It was actively to be eliminated. You know we've never seen that in the history of art. It is not part of modernism. It is not part of the history of western art. It was a re-sponse, in my mind, to a very specific political situation, mainly censorship because of Cold War issues in this country. I feel really embarrassed when I see that ideological view now adopted by contemporary European artists and critics and curators. I feel that this is the most obvious manifestation of American imperialism that one could ask for, and people are buying it whole-hog, which I think is very, very dangerous. My feeling is that the recovery of political and social content in this country during the last twenty years, first of all has been made with a great deal of effort and fighting against a great deal of resistance. And I think it is because of the censorship about content

during the fifties that we now think that somehow there is discontinuity in the development of western art. And that somehow we are beyond modernism and we are into a completely different age—virtual reality and all of that. I mean the continuities are so clear if we understand that…if we simply isolate that *one* period in the history of American art as basically a kind of a fluke, an idiosyncrasy of history that we need to understand in context and get past. So I feel that the real continuity with the history of art in Europe and in the United States is connecting the political content and concerns that artists are showing now with the very similar concerns and commitments of artists in the past. And to me, I do not see that as completely different, I see it as continuous with the history of art.

## Jaume Xifra

All this aspect is in full mutation, as is our future. I think that art is increasingly taking a more didactic and social aspect. Besides, the media are growing relentlessly and art is acquiring an interest…I mean, it always has this interest as a collector's item, but more and more art is now acquiring a social interest. I mean that any reflection, whatever it is, is based on—art media reflections, or on the contents of these media—it gives the art piece a more informative dimension rather than a value as a collector's or a museum object. I mean, it ends up going to a collection or a museum, because it is the witness of a cultural gain. But it seems to me that this mutation, this movement, lies in the fact that media, galleries, museums, are getting structured more to offer information not to collectors but to society in general, to enrich society at large.

## Dan Graham

I think it is when people were interested in art as part of a more general culture, writing, cinema, whatever, philosophy…I think that was more part of the later sixties and early seventies; also, artists themselves saw their work that way. And I am not sure, now that art is much more valuable, much more expensive, and specialization has also come into the situation again, that you can view it in the same terms. I think, economically the eighties changed everything because art became…Let us say, in the sixties when art was connected to money, it was the idea that money was to be thrown away or be replenished. It didn't matter, just like things could be burned and didn't have to be saved. So you had all kinds of people who were interested in art and if they really found it…If their interests were superficial, they wouldn't stay there: they would get disinterested and go on to something else. You didn't have people building empires, or empires that would last in any way. The thing about art that was good was that it was protean, that you did not know what was coming along. It was connected to fashion, a little bit like rock music, but on the other hand, you could still build amazing things inside of that structure. I think today people have got more conservative about the economic aspect, probably because they knew that these things…you could get money anywhere, an idea…and that you did not have to protect existing situations. In other words, I think that art became more like real estate, and the investment that you had in art or a particular artist became very important. I think that has been more of the eighties phenomena.

**Hans Haacke**
In the eighties possibly more than before (but it was never absent) art was for all involved—artists, dealers, museum people, collectors and whoever had a stake in it, including public officials…Among other things it was a means to gain social prestige. A good number of people who made money very quickly, lots of money, had a need to surround themselves with things that were generally perceived as disinterested, as not connected with the low marketplace. And they surrounded themselves with art. While at the same time, in a somewhat schizophrenic way, of course, they were also intrigued. And, in fact, they were interested in the increase of the monetary value of the art that they collected. If art does not lend itself to social prestige as much as it did before, then perhaps there will be a slight shift, but which way it may be going, I don't know. But that makes me think of another aspect that we have not talked about: the corporate role that became very strong in the seventies and the eighties. Corporations got involved in the arts because the arts were something with which they could polish their image, and through the arts, they also become more forceful in their lobbying for interests very close to the bottom line. If the arts loose some of the glamour that they

had, then it is quite conceivable that corporations will also withdraw some or all of their support, because they have to justify to their shareholders the expenses that they incur in supporting art. And the shareholders, of course, are interested in the bottom line, not in the aura.

**Allan Kaprow**
Money by itself is neutral. It is who has the money and the values of who has most of the money that make the world go round. Sometimes make it stop. Let me start this over again, that is too metaphorical. Artwork, when it becomes expensive, when it becomes a precious object that sells for a lot of money, is associated with power, because those who have the money normally have the political power in this world. That is the problem with the kind of association that a lot of artists have with success. If you become successful in terms of the economic system that art is part of, you become automatically bound up with the value systems, the value range, of those who have the money to have leisure time and to purchase artwork. This is a problem for a lot of us, because we don't all share the values of those who have power. It's very difficult for an artist, therefore, to choose an alternative market, that is to choose a market which will allow that artist to survive in the way that a plumber or a carpenter survives—to earn a living. There is no other market other than the luxury trade, and for a while it seemed that the museum could provide—the public museum, not the private museum—could provide that alternative, because it was using public monies, the taxpayers' money. However, those who are the directors of such museums come out of the same rather patrician tradition that the private collector comes out of, and they're, in fact, symbiotically related because many donations from such collectors go directly to the public museums and the association of artists with that world becomes absolute. There is no way out of it. I worried about that right from the beginning. A lot of artists worried about it but felt that they could not do anything about it and therefore decided in effect to join the system. My choice was a different one. To speak very frankly, I decided to get a job which would leave me absolutely free of the commercial system in the art world. So I could do whatever I felt was necessary in my work, without depending

upon sales. Now that is not the choice of other artists, so I cannot speak for them. But it allows me to look at the art world—at dealers, at museum people, at critics, at collectors—with a certain degree of detachment, without personal involvement and, therefore, I think, with a certain amount of sympathy for the problems which everybody is sharing.

### Braco Dimitrijevich

I see the nineties as a time of interesting individuals and not a time of movements. So I think we are going to face a certain pluralism which was not evident in the eighties or in the late seventies. What is very important for myself as an individual is to appreciate a variety of concepts that exist around me, but to approach all that with great subjectivity, with a maximum of tolerance—as opposed to trying to associate myself with other individuals who are dealing with similar concepts. That is how we might achieve a certain richness in this field of activity.

### Adrian Piper

Places like museums and galleries can be incredibly important if they are doing their job right, and if they are not, that is terrible, that is very bad, and that is why I am so against the kind of "spectacle presentation" that we see in museums now, because when that happens, it frightens people off. It sends them back to the TV.

### Krzystof Wodiczko

I think that most of the artists and curators and critics and directors of the museums and owners of the galleries and all the experts and teachers…Most of them, it seems to me, are absolutely unprepared to continue the discussion— theoretically and also in terms of practice, aesthetic practice—about the ways of engagement and work in life, in contemporary societies and environments, with the use of contemporary media and means of address, and also in relation to the production of meaning and the powerful aesthetic practice that is being imposed on us or accepted by us, is being absorbed half-consciously or con- sciously by us. There is this fantastic work of art which is the environment itself: cities and real estate, architecture and state, architecture which barricades and organizes and structures our life and our consciousness in this contemporary postindustrial world.

**Hans Haacke**
Well, since artists have gained the public status that they have, and general readership publications and television report about what is happening in the arts—irrespective of what they do, they participate, as I said earlier, in the shaping of our general collective consciousness. They participate in molding public opinion, whether they are intending to do so or not. What happens in the big cities, I mean only in the western world—New York, London, Paris, German cities and so forth—and how that is translated through various channels (the press, the media and so forth) into the general consciousness, that is also where individual artists through their work add strands to the fabric of how we see ourselves today and what values we have and what values we might want to be allied with. And so in a very indirect way, but not a negligible way I might add, the arts have a social impact, and one should not underestimate it. It is part of what people might have called "the consciousness industry." And if you look at the size of the art world in terms of the money that is being transacted compared to the other parts of the "consciousness industry," it is minuscule. But if you look at what happens in this small sector, how that rubs off on the rest of it, it is astounding. And that is something that we should not lose sight of. And therefore, it would be good if artists would think about not only what are they going to sell in the next show, whether they're going to be invited to *Documenta* or such things, but also what their work may contribute to the general fabric of public opinion.

# List of Participants

Interviews completed between 1983 and 1991.*

### Chapter 1

The Dealers

Lucio Amelio, Naples. Daniel Templon, Paris. Ronald Feldman, New York. Leo Castelli, New York. Richard Kuhlenschmidt, Los Angeles. Ivan Karp, New York. Holly Solomon, New York.

### Chapter 2

The Collectors

Acey and Bill Wolgin, Philadelphia. Eric and Sylvie Boissonas, Paris. Bob Calle, Paris. Isabel de Pedro and Raphael Tous, Barcelona. Gianni Rampa, Naples. Marcia Weisman, Los Angeles. Robert Rowan, Los Angeles. Giuseppe Panza di Biumo, Varese. Herman Daled, Brussels. Toshio Ohara, Tokyo. Fernando Vijande, Madrid.

### Chapter 3

The Galleries

Sonnabend, Ileana Sonnabend, New York. Diacono, Mario Diacono, Rome. Stadler, Rodolphe Stadler, Paris. Durand-Dessert, Michel Durand-Dessert, Paris. Joan Prats, Joan de Muga, Barcelona. Ciento, Marisa Díez, Barcelona. Glenn Lewis, Vancouver. Marian Goodman, Marian Goodman, New York. Oil and Steel, Richard Bellamy, New York. Rosamund Felsen, Rosamund Felsen, Los Angeles. L.A.C.E., Joy Silverman, Los Angeles. Mary Boone, Mary Boone, New York. Metro Pictures, Helene Winer, New York. Otis Parson Gallery, Al Nodal, Los Angeles.

### Chapter 4

The Museum

Dominique Bozo, Centre Pompidou, Paris. Rudi Fuchs, Van Abbe Museum, Eindhoven. Wolfgang Becker, Neue Gallery, Aachen. Jean-Christophe Amman, Basler Kunstverein, Basel. Margit Rowell, Centre Pompidou, Paris. Johannes Cladders, Mönchengladbach Museum, Monchengladbach. Evelyn Weiss, W.R. Museum, Cologne. Martin Kunz, Kunstmuseum Luzern, Lucerne. Harald Szeeman, Zurich. Pontus Hulten, Paris. Lola Bonora, Palazzo dei Diamanti, Ferrara. Josep Ainud de Lasarte, Muse d'Art de Catalunya, Barcelona. Anne D'Haroncourt, Philadephia Museum, Philadelphia. Daivd Ross, Institute of Contemporary Art, Boston. Kathy Halbreich, Hayden Gallery– M.I.T., Cambridge. Richard Armstrong, Whitney Museum, New York. Riva Caslteman, Museum of Modern Art, New York. Marc Schepps, Tel Aviv Museum of Art, Tel Aviv. Yona Fischer, The Israel Museum, Jerusalem. Richard Koshalek, Museum of Contemporary Art, Los Angeles. Diane Shamash, Santa Barbara Museum, Santa Barbara. Thomas Messer, The Guggenheim Museum, New York. Marcia Tucker, The New Museum of Contemporary Art, New York. Gary Garrels, DIA Art Foundation, New York. Carmen Giménez, The Guggenheim Museum, New York. María Corral, Centro de Arte Reina Sofia, Madrid. Jean-Hubert Martin, Paris. Yves Gevaert, Brussels. Kasper König, Porticus, Frankfurt. Fumio Nanjo, Tokyo. Jean-Louis Froment, capcMusee d'art contemporain, Bordeaux.

### Chapter 5

The Guides

Ellen Bretman, Kathleen Byrnes, Kathy Dealp, Newport Harbor Museum. Mary Drobny, Jan Laney, Barbara Stewart, Claudia Wishnow, Long Beach Museum of Art.

### Chapter 6
The Critics
Bernard Marcade, Paris. Pierre Restany, Paris. Filiberto Menna, Rome. Achille Bonito-Oliva, Rome. Nina Dimitrijevic, London. Guy Brett, London. Thomas Wulffen, Berlin. Yves Michaud, Paris. Tommaso Trini, Milan. Lorenzo Mango, Rome. Lucy Lippard, New York. Peter Frank, New York. Catherine Strasser, Paris. Bernard Lamarche-Vadel, Paris. Catherine Millet, Paris. Maria Luisa Borras, Barcelona. Daniel Giralt-Miracle, Barcelona. Victoria Combalía, Barcelona. Christopher Knight, Los Angeles. Donald Kuspit, New York. Craig Owens, New York. Benjamin Buchloh, New York. Dore Ashton, New York. Jeanne Randolph, Toronto.

### Chapter 7
The Media
Robert Atkins, New York. Shelly Rice, New York. Heidi Grundman, Vienna. Giancarlo Politi, *Flash Art,* Milan. François Pluchart, *L'Art Vivant,* Paris. Geneviève Breerette, *Le Monde,* Paris. Teri Wehn Damisch, R.T.F., Paris. Willy Bongard, *Art Aktuell,* Cologne. Josep Iglesias del Marquet. *La Vanguardia*, Barcelona. Rosa Queralt, Radio 4, Barcelona. Roman Gubern, Universidad Autonoma, Barcelona. Rene Berger, University of Lausanne, Lausanne. Judith Hoffberg, *Umbrella*, Los Angeles. Michael Brenson, *New York Times*, New York. Jef Cornelis, R.T.B., Brussels. Mary-Anne Staniszeski, New York. Paul Taylor, New York. Brian Wallis, *Art in America,* New York. Tatsumi Shinoda, Mizue, Tokyo. David Antin, San Diego.

### Chapter 8- Epilogue
Luciano Fabro, Milan. Joseph Beuys, Dusseldorf. Fernando de Fillipi, Milan. Jaum Xifra, Paris. Daniel Buren, Paris. John Baldessari, Los Angeles. Alan Kaprow, San Diego. Braco Dimitrijevic, London. Krzysztof Wodiczko, New York. Dan Graham, New York. Hans Haacke, New York. Regina Silveira, Sao Paulo. Adrian Piper, Wellesley.

*The locations and institutional affiliations listed reflect the time when the interviews were conducted.

# Credits

**Camera by:**
Caterina Borelli
*and*
Kate Craig, Steve O'Rear, Ted Hardin, Art Nomua

**Production and Post -Production**
Bruce Tovskyltrit Batsry
Ted Hardin
Walter Zooi
Joe Leonardi
Jayce Salloum
Philippe Lavaill
Alessandro Melodia
Josep Maria Vilardosa
John Barnett
Jordi Agusti
Jordi Torrent
Aleix Gallardet
Doug Brown
Luis Nicolau
Hank Bull
Larry Loganbill
Elizabeth Vander Zaag
Tim V. S. Westbury
Barnaby Levy

**Executive Producer**
Caterina Borelli

With the collaboration of:
C.A.V.S.–E.V.R., Massachusetts Institute of Technology, Cambridge
Visual Arts–Media Studies, Scripps Institute of Oceanography,
    University of California, San Diego
Honolulu Academy of Art
Radio, Cinema and Television Dept., Temple University, Philadelphia
Canal Video, Videografia, Barcelona
Electronic Arts Intermix, New York
The Station, Long Beach Museum of Art
Art Studio–Media Arts, The Banff Centre, Banff
Wexner Center for the Arts, The Ohio State University, Columbus

**Special Thanks to:**
Kathy Huffman
Lorne Falk
Lory Zippay
Melodie Calvert
Martine Bour
Antoni Mercader

*and*
Bill Horrigan, Mireia Sentis, Chip Lord, Danielle Tilkin, Susan Harris, Fern Bayer, Barbara London, Kira Perov, Manami Fujimori, Steven Vitiello, Marshall Reeese, Joan Rabascall, Jean-Luc Saumaude, Enrique Riu, Christine Courreges, H. Goed, Giovanna Zamboni, Maia Giacobbe Borelli, Peter d'Agostino, Angel Shaw, Sherman George, Doug Walla, James R. Blattenberger

*and*
United McGill Corp., Columbus; Columbus City Center, Columbus; Nationwide One Plaza, Columbus; Caltrans Los Angeles Freeway; International House of Japan; the Guides of the Long Beach Museum of Art and the Newport Museum Harbor Museum

*and*
the particpants in the project *Between the Frames*

**Partially funded by:**
The John Simon Guggenheim Memorial Foundation, New York
The New York State Council on the Arts
Generalitat de Catalunya, Department de Cultura-Serveis
    Audiovisuals, Barcelona
Long Beach Museum–Video Program, Long Beach
Commandes de l Etat, Centre National des Arts Plastiques, Paris